UNIVERSITY OF CALIFORNIA,
BERKELEY

PHYSICS PROBLEMS

With Solutions

UNIVERSITY OF CALIFORNIA, BERKELEY

PHYSICS PROBLEMS

With Solutions

MIN CHEN
Massachusetts Institute of Technology

Prentice-Hall, Inc., *Englewood Cliffs, New Jersey*

Library of Congress Cataloging in Publication Data

Chen, Min, 1940-
University of California, Berkeley, physics problems, with solutions.

1. Physics--Problems, exercises, etc. I. Title.
QC32.C43 530'.076 74-8798
ISBN 0-13-938902-4

10 9 8 7 6 5 4 3 2 1

Printed in the United States of America

PRENTICE-HALL INTERNATIONAL, INC., LONDON
PRENTICE-HALL OF AUSTRALIA, PTY. LTD., SYDNEY
PRENTICE-HALL OF CANADA, LTD., TORONTO
PRENTICE-HALL OF INDIA PRIVATE LIMITED, NEW DELHI
PRENTICE-HALL OF JAPAN, INC., TOKYO

CONTENTS

PREFACE vii

MECHANICS PROBLEMS 1

Solution Set of Mechanics Problems 24

ELECTRICITY AND MAGNETISM 99

Solution Set of Electricity and Magnetism 119

HEAT, STATISTICAL MECHANICS, AND OPTICS 182

Solution Set of Heat, Statistical Mechanics, and Optics 206

ATOMIC PHYSICS AND QUANTUM MECHANICS 263

Solution Set of Atomic Physics and Quantum Mechanics 288

APPENDIX 351

REFERENCES 354

PREFACE

The problems in this book were selected from the written preliminary examinations given at the University of California at Berkeley from 1959 to 1968. They were developed by faculty members of the Physics Department at Berkeley who served on the successive preliminary examinations committees during the past nine years. The examinations are given twice a year. All entering graduate students are required to take this two-day comprehensive examination covering the subject matters of those undergraduate courses that deal with the fields of Mechanics, Electricity and Magnetism, Optics, Heat and Atomic Physics. The examination is divided into four subsections, each three hours in length. No notes, tables or other references are to be used. Since these problems were submitted for publication, one major change has been instituted: a slightly more advanced knowledge of quantum mechanics is now required.

This examination serves two purposes. It is a placement test to see whether entering students need additional study in certain areas of undergraduate physics, and it is used as a qualification test for advanced degrees. Students who perform successfully proceed directly to the oral preliminary examination. Students are allowed three chances to pass the written examination: once as a placement test and twice, if necessary, as a test of their qualification for advanced degrees. Failure to pass the preliminary examination on the second attempt would disqualify a student from the Ph. D. program at the University of California, Berkeley.

The selected problems are put in subsections similar to the original form of the preliminary examinations. Each problem is marked by a certain weight (points) and the total points in each subsection add up to one hundred points. One should take no more than three hours to finish each subsection. To use this book most advantageously, readers should first try to solve these problems by themselves without referring to the solutions nor to any other books.

We think that prospective students at Berkeley and elsewhere might like a collection of the problems given, with their solutions. Such a collection could help materially in their preparation for the test and more broadly in gauging their proficiency. Study with a friend or two, and related "bull sessions" might be an amusing way of preparing.

In the appendix, we have added a list of books containing the basic information the student will need to fill any deficiency in his preparation. Often, however, the text one has studied is the best source of information.

The solutions were prepared by the author and reviewed by Drs. C. Fong, R. Kunselman, K. Wang, C. F. Chan, and Chin-fu-Tsang. Although they were carefully prepared and checked by the reviewers, errors are still inevitable. Comments are extremely welcome by the author. The Physics Department of the University of California at Berkeley takes no responsibility for their correctness.

The author is very grateful to Professor Emilio Segrè for his encouragement and valuable advice. He would also like to thank the reviewers for their faithful work and numerous suggestions and especially Suzanne, for her patient preparation of the manuscript.

MIN CHEN

UNIVERSITY OF CALIFORNIA,

BERKELEY

PHYSICS PROBLEMS

With Solutions

MECHANICS PROBLEMS

(I-1) (15 points) A particle of mass m_1, momentum P_1, collides elastically with a particle of mass m_2 which is initially at rest. Calculate the maximum possible momentum in the laboratory system of particle m_2 after the collision. (Use relativistically correct formulas.) Apply your result to the case of a proton having a momentum equal to its rest energy/c colliding with an electron at rest. Give a numerical value for the maximum momentum of the electron after the collision in MeV/c.

(I-2) (15 points) A uniform thin rod of length L, mass m, is lying on the usual smooth horizontal table. A horizontal impulse, $\hat{F}$, is suddenly applied perpendicular to the rod at one end.

(a) How far does the rod travel during the time it takes to make one complete revolution?

(b) What are the translational, rotational, and total kinetic energies of the rod after the impulse?

(I-3) (15 points) A uniform bar of length L and mass m is supported at its ends by identical springs with elastic constant k. Motion is initiated by depressing one end by a small distance a and releasing it from rest. Solve the problem of the motion, identifying normal modes and frequencies. Make a sketch indicating the normal modes.

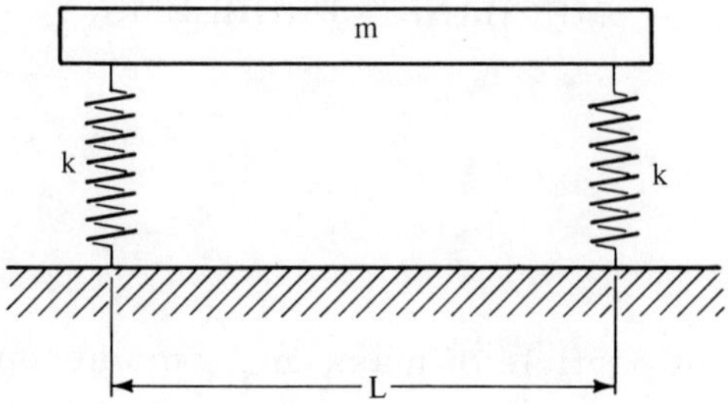

(I-4) (10 points) Consider a real $N \times N$ matrix A. If A is symmetric, $A_{ij} = A_{ji}$, prove that all its eigenvalues are real. If A is not symmetric, prove that any nonreal eigenvalues must occur in complex-conjugate pairs.

(I-5) (15 points) A body of mass m is dropped from rest from a height h at latitude 45° in the northern hemisphere ($h \ll$ radius of earth). Where will it land relative to a plumb bob suspended from the point of release? Be sure to state both the magnitude and direction of the displacement.

(I-6) (15 points) A flexible string of length L is stretched with tension T between fixed supports. Its mass per unit length is ρ, so that its total mass is $M = \rho L$. The string is set into vibration with a hammer blow which impulsively imparts a transverse velocity v_0 to a small segment of length a at the center. Evaluate the amplitudes of the lowest three harmonics excited.

(I-7) (5 points) For a solid body moving through the air at a very high velocity $\bar{V}$ (faster than the kinetic velocity of the air molecules), show that the drag force is proportional to $A\bar{V}^2$, where A is the frontal area of the body.

(I-8) (5 points) What would be the length of one day if the earth spun so fast that bodies floated at the equator?

(I-9) (5 points) Define cyclic coordinate. Why is it useful in solving problems?

(II-1) (10 points)

(a) (5 points) Show that the only possible real eigenvalue of a skew Hermitian matrix (i.e., $A^{+} = -A$) is zero.

(b) (5 points) Evaluate $\int_{-\infty}^{\infty} \frac{\sin^2 x}{x^2}\, dx$ using the method of residues.

(II-2) (15 points)

(a) (10 points) How much energy would have to be supplied to break up the earth completely (i.e., remove all mass to ∞)?

(b) (5 points) If the earth were then reassembled rapidly (i.e., so fast that no energy is radiated away), would it melt? Justify your answer.

(II-3) (30 points)

(a) (20 points) A rigid rod of length a and mass m is suspended by equal massless threads of length L fastened by its ends. While hanging at rest, it receives a small impulse $\hat{P}$ at one end in a direction perpendicular to the rod and to the thread. Calculate the normal frequencies and the amplitudes of the associated normal modes in the subsequent motion.

(b) (10 points) If the rod is hanging at rest and suddenly one of the strings is cut, what is the tension in the other string immediately thereafter?

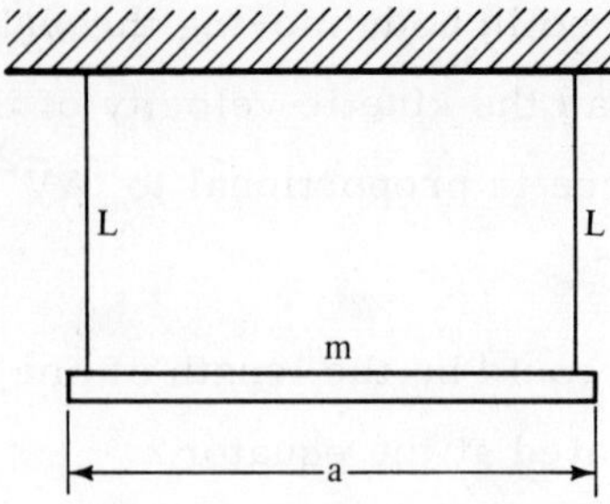

(II-4) (25 points) A particle of mass m moves under the action of a central force whose potential is $V(r) = Kr^3$ $(K > 0)$.

(a) (10 points) For what energy and angular momentum will the orbit be a circle of radius a about the origin?

(b) (5 points) What is the period of this circular motion?

(c) (10 points) If the particle is slightly disturbed from this circular motion, what will be the period of small radial oscillations about $r = a$?

(II-5) (20 points) It has been suggested that monoenergetic photon beams of high energy can be obtained by scattering laser beams from energetic electrons produced in electron accelerators. Derive a formula for the maximum energy of the scattered photons in terms of the energy of the laser photons and the energy of the electrons in the beam. Make a numerical estimate for the case of a Ruby laser beam which is scattered from the 20 GeV electrons available at the Stanford Linear Accelerator.

(III-1) (15 points) A marble of mass M and radius a rolls without slipping on an inclined plane making an angle δ with the horizontal.

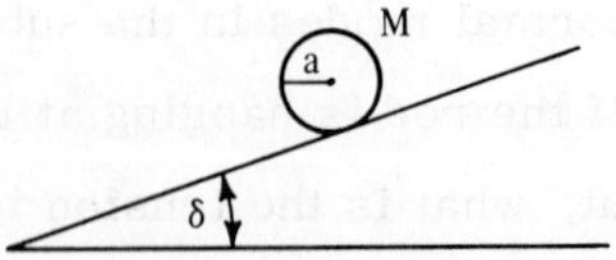

(a) Calculate the acceleration of the center of the marble.

(b) If the marble is started with initial velocity v_0 directly up the inclined plane, how long will it be before the marble returns to its starting point?

(III-2) (15 points) An unstable particle of rest energy 1000 MeV decays into a μ-meson (m_0c^2 = 100 MeV) and a neutrino with a mean life, when at rest, of 10^{-8} sec.

(a) Calculate the mean decay distance when the particle has a momentum of 1000 MeV/c.

(b) What is the energy of the μ-meson if it is emitted at an angle 15° ?

(III-3) (15 points) Consider two rigid rods of length L_1 and L_2 and masses M_1 and M_2 fixed to a horizontal massless rod with torsion constant K which is free to rotate about its longitudinal axis only. The massless rod is supported at both ends, and the system hangs under the influence of gravity. Write the Lagrangian for the system (define your variables) and find the equation of motion.

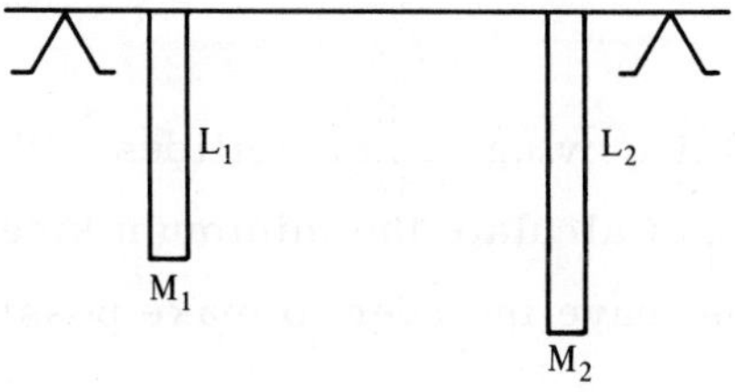

(III-4) (25 points) A particle is placed on a rough plane inclined at the angle θ, where tan θ = μ = coefficient of friction (both static and dynamic). A string attached to the particle passes through a small hole in the plane. The string is pulled so slowly that you may consider the particle to be in static equilibrium at all times. Find the path of the particle on the inclined plane.

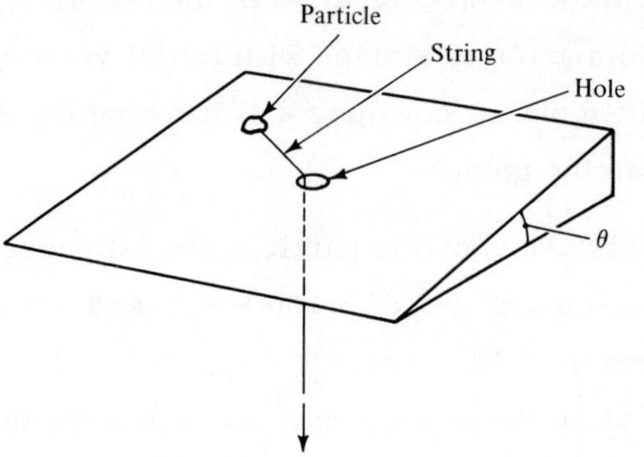

(III-5) (15 points) A bar of mass M and length L is free to move in a vertical plane, as shown below:

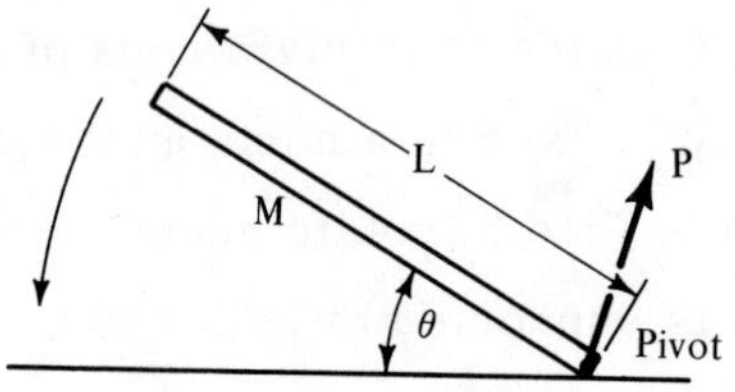

If the bar starts from rest at an angle of 30° above the horizontal, calculate the force on the pivot when it swings through the horizontal direction.

(III-6) (15 points) A moving proton collides with another proton which is initially at rest. Calculate the minimum kinetic energy that the moving proton must have in order to make possible the reaction

$$p + p = p + p + p + \overline{p}$$

The antiproton $(\overline{p})$ has the same mass as the proton. Express your result as the ratio of the kinetic energy to the rest energy of the proton. What is the approximate numerical value for the kinetic energy in MeV?

(IV-1) (15 points) Assume that the tidal waves on the earth are caused by the sun only.

(a) Where does the energy dissipated in the tides come from, and what is the maximum energy available?

(b) How is the total angular momentum of the system conserved?

(IV-2) (5 points) The life time of the μ-meson is 2×10^{-6} sec. A beam of μ-mesons emerges from a cyclotron with velocity 0.8c, where c = velocity of light. What would be the mean life of the μ-mesons in this beam as observed in the laboratory?

(IV-3) (5 points) In the figure, a pulley of negligible weight is suspended by a spring balance. Weights of 1 kg and 5 kg, respectively, are attached to opposite ends of a string passing over the pulley and move with acceleration because of gravity. During

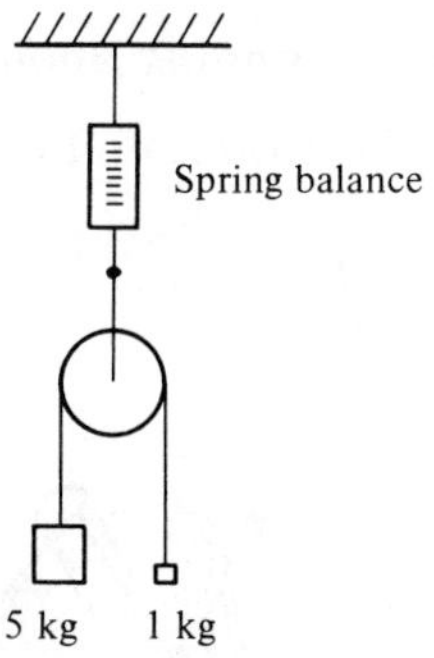

their motion, will the spring balance read a weight of 6 kg, less than 6 kg, or more than 6 kg?

(IV-4) (5 points) A bullet is fired from a rifle. If the rifle were allowed to recoil freely (i.e., without being restrained by a person's shoulder) its kinetic energy as a result of recoil would be (equal to) (less than) (greater than) that of the bullet.

(IV-5) (5 points) The shaded area in the figure represents an L-shaped uniform plate. What are the x and y coordinates of its center of mass?

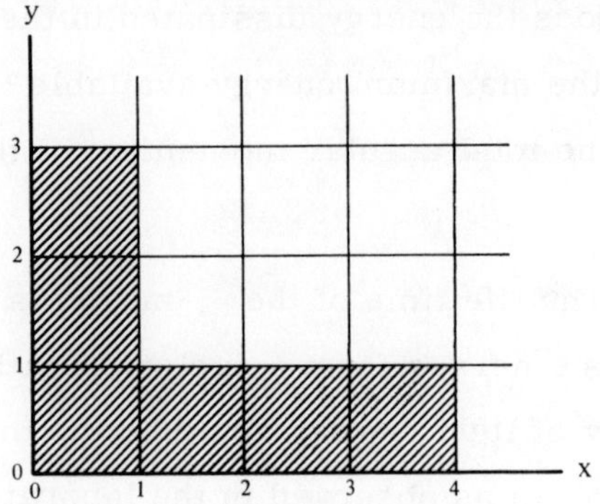

$\bar{x}$ = __________ $\bar{y}$ = __________

(IV-6) (5 points) An ice cube is placed in a glass of water at 0° C. After the ice has melted, the level of water in the glass has (risen) (fallen) (remained the same).

(IV-7) (5 points) The top is spinning about its axis in the sense indicated by the arrow. The lower end of the top pivots on a table. Does the top, as seen from above looking down upon it, precess clockwise or counterclockwise?

(IV-8) (5 points) Masses m and 3m are attached to the two ends of a spring of spring constant k. What is the period of oscillation?

(IV-9) (5 points) The mutual potential energy V of two particles depends on their mutual distance r as follows:

$$V = \frac{a}{r^2} - \frac{b}{r}$$

where a and b are positive constants. For what separation r are the particles in static equilibrium?

(IV-10) (5 points) Two particles of equal mass are connected by springs as shown and are free to execute longitudinal one-dimensional oscillations. Indicate by arrows (or describe otherwise) the relative magnitudes and phases of the particle displacements corresponding to the two normal modes.

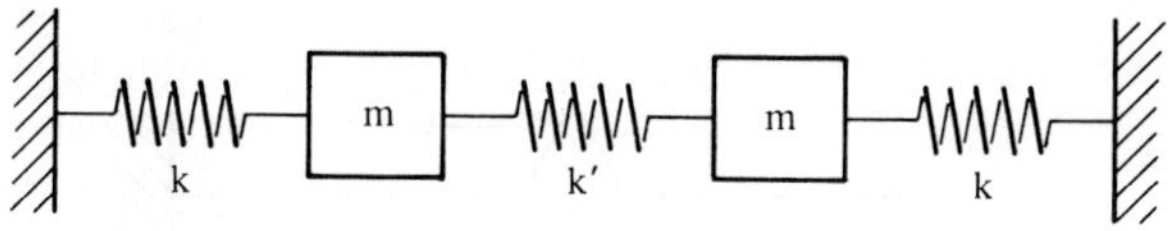

(a) Mode No. 1:

(b) Mode No. 2:

(IV-11) (5 points) Suppose that the radius of the earth were to shrink by 1%, its mass remaining the same. Would the acceleration due to gravity g on the earth's surface increase or decrease, and by what percentage?

(IV-12) (5 points) An empty cylindrical tin can and a can tightly packed with luncheon meat both start rolling down an inclined plane at the same instant. Which can reaches the bottom first?

(IV-13) (5 points) A cylindrical can filled with water to a height of 40 cm has on its side two small holes of equal area, one at a height of 10 cm and the other at a height of 30 cm. At the initial time, what is

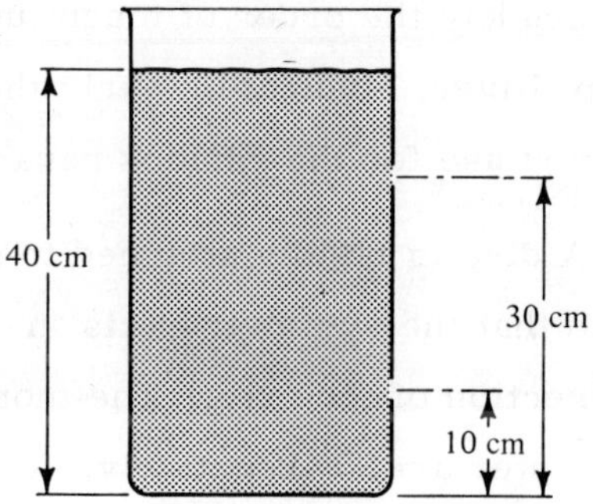

the ratio of the mass of water flowing per second through the lower hole to that flowing through the upper hole?

(IV-14) (5 points) A long taut string has its right end fastened to a rigid wall. A transverse disturbance in the form of an isosceles triangle propagates with a velocity of 1 m/sec towards the wall. At some initial time, the center of the triangular pulse is 2 m to the left of the wall. Draw the shape of the string after two seconds <u>and</u> after five seconds.

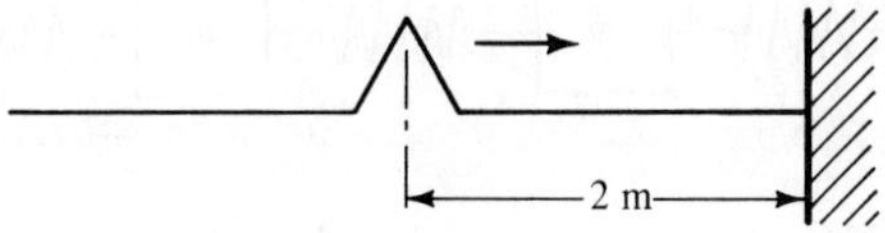

(a) After two seconds:

(b) After five seconds:

(IV-15) (10 points) The formula for the Coriolis force is
$\vec{F} = -2m\vec{\omega} \times \vec{v}$.

(a) In what situation does the formula apply and what do the symbols mean?

(b) A river runs southward along a channel of width W in north latitude λ. Then the water level on the east bank of the river will be different from the level on the west bank. Which will be higher?

(c) Find the difference in height, carefully defining any symbols used.

(d) Estimate crudely the order of magnitude of this effect for the Mississippi River. Indicate clearly the estimates (or guesses) that you use for the various parameters.

(IV-16) (10 points) A disc is rigidly attached to an axle passing through its center so that the symmetry axis $\hat{n}$ of the disc makes an angle δ with the direction of the axle. The moments of inertia of the disc relative to its center are, respectively, C about the symmetry axis $\hat{n}$ and A about any direction $\hat{n}'$ perpendicular to $\hat{n}$. The axle spins with constant angular velocity ω. Find the magnitude of the torque exerted on the bearings supporting the axle.

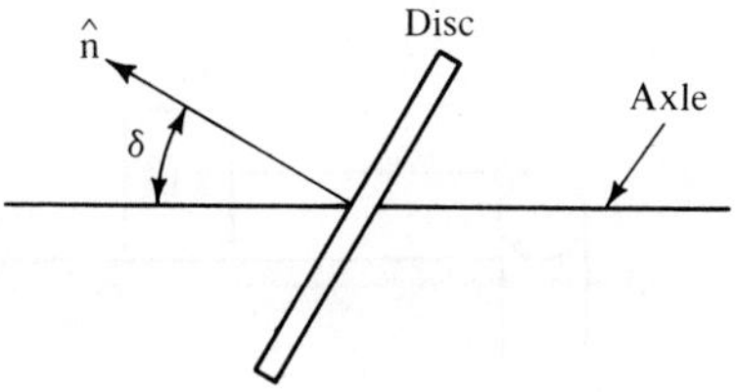

(V-1) (10 points) A block of mass M rests on a frictionless horizontal table and is connected to two fixed posts by springs having spring constants k_1 and k_2 respectively.

(a) If the block is displaced slightly from its equilibrium position, what is the frequency of vibration? Suppose that the block is vibrating with amplitude A and that, at the instant that it is passing through its equilibrium position, a mass m is dropped vertically onto the block and sticks to it. Find:

(b) the new frequency of vibration;

(c) the new amplitude of vibration.

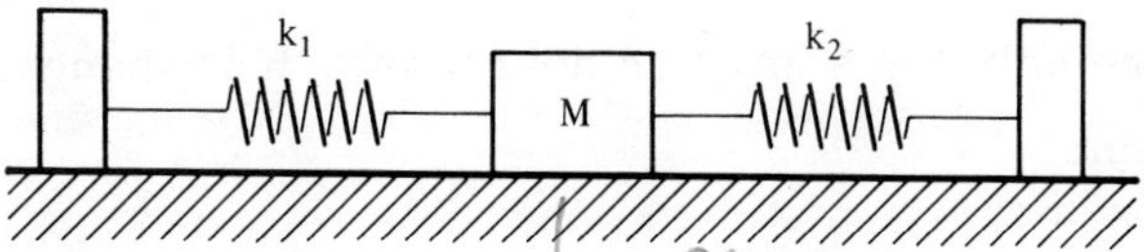

(V-2) (10 points) A wedge of mass M, whose faces form an angle β with each other, lies on a smooth horizontal table. A block of mass m is placed on the wedge and is allowed to slide down. Neglect all friction. What is the acceleration of the wedge along the table (before the block reaches the table)?

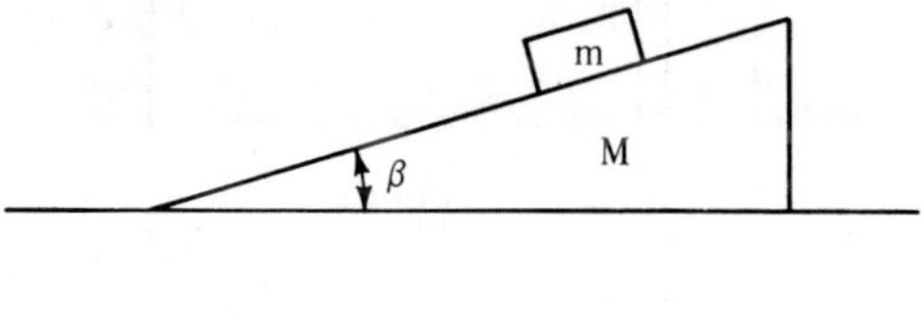

(V-3) (3 points) In the figure a frictionless, massless pulley is fixed to the edge of a frictionless table. The two blocks of mass 10 kg and 5 kg are connected by a weightless string passing over the pulley. Is the tension in the string greater than, equal to, or less than it would be if the 5 kg block were glued down?

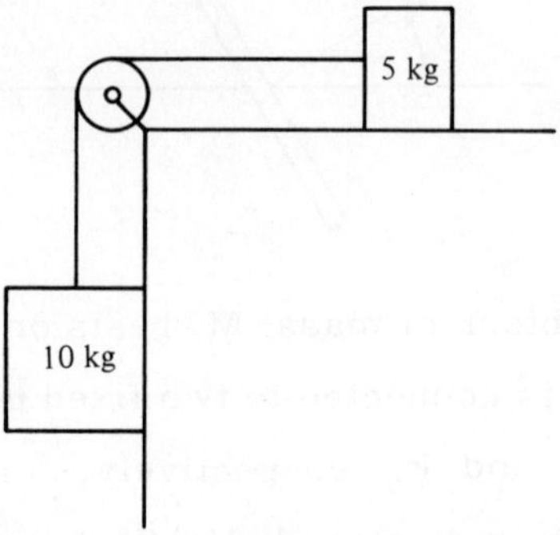

(V-4) (3 points) A circular disc is rotating about its center in the horizontal plane. A razor blade is balanced on edge in a groove along a radius of the wheel. If the blade is pulled along the radius toward the center of the wheel, will it tend to fall in the direction of rotation or opposite to the direction of rotation?

(V-5) (3 points) Imagine the radius of the earth shrinking by 1%, its mass remaining the same. Would the kinetic energy of rotation increase, remain the same, or decrease? If it changes, by how many percent?

(V-6) (3 points) A uniform thin rod of mass M and length L hangs from a frictionless pivot and is connected at the bottom by a spring to the wall as shown. The spring constant is K. What is the period of the motion?

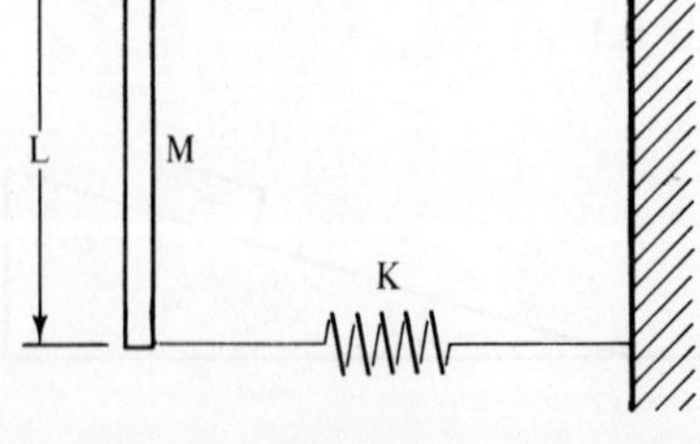

(V-7) (3 points) Two thin rods with mass M and length L are hung from pivots and are connected together at the bottom by a spring such that at rest the rods hang straight down. The spring constant is K. Indicate by arrows the nature of the oscillations in the two normal modes. What are the frequencies for these two normal modes?

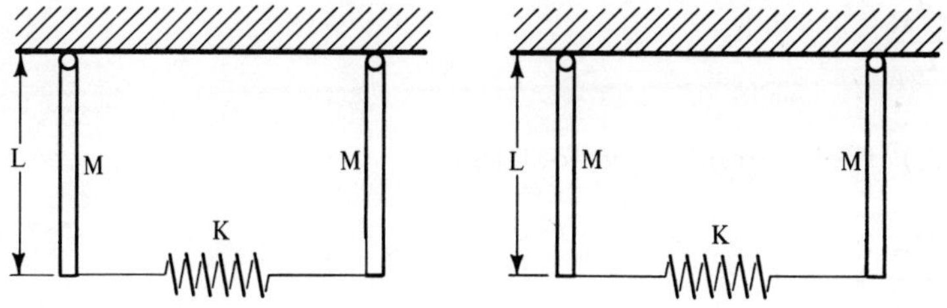

Mode I. ______________ Mode II. ______________

(V-8) (3 points) What is the ratio of the mass of the sun to the mass of the earth to one significant figure?

(V-9) (3 points) Two persons of equal weight are hanging by their hands from the ends of a rope hung over a frictionless pulley. They begin to climb. One person can climb twice the speed of the other (with respect to the rope). Who gets to the top first? (a) the faster climber, (b) the slower climber, (c) get there together, or (d) indeterminate.

(V-10) (3 points) A cork is submerged in a pail of water by a spring attached to the bottom of a pail. The pail is held by a child in an elevator. During the initial acceleration as the elevator travels to the next lower floor, will the displacement of the spring (a) increase, (b) decrease, (c) remain the same?

(V-11) (3 points) A satellite is launched into a circular orbit of radius R. A second satellite is launched into an orbit of diameter 1.01R. Is its period larger, the same, or smaller? If its period is different, by how many percent?

(V-12) (3 points) A sphere of radius R is released in a liquid of viscosity η, so that by Stokes' law its drag is $6\pi\eta Rv$. Simultaneously a second sphere of identical mass but with radius 2R is released. What is the ratio of:

(a) their initial accelerations?

$$\frac{a_R}{a_{2R}} =$$

(b) their terminal velocities?

$$\frac{v_R}{v_{2R}} =$$

(V-13) (3 points) A particle moves in the field of force given by:

(1) $F_x = 2yz(1 - 6xyz)$, $F_y = 2xz(1 - 6xyz)$, $F_z = 2xy(1 - 6xyz)$.

(2) $F_x = y^2 + z^2 + 2(xy + yz + zx)$, $F_y = x^2 + z^2 + 2(xy + yz + zx)$,

$F_z = x^2 + y^2 + 2(xy + yz + zx)$.

(a) For which forces may a potential function $V(x, y, z)$ be defined?

(b) For which forces will the total energy (kinetic plus potential) be constant during the motion of the particle?

(V-14) (20 points) A sphere of radius r is initially spinning around its axis with angular velocity ω. The axis is horizontal. When dropped on a surface with a coefficient of friction μ, it slips and then rolls without slipping.

(a) What is the final linear velocity of the center of mass?

(b) What is the distance it travels before achieving this velocity?

(c) Repeat the calculation of (a) in the case that μ is so large that there is no initial slip.

(V-15) (13 points) A football in the shape of an ellipsoid of revolution has moments of inertia about its principal axes of I_1, I_2, and I_3. When thrown it is accidentally given an angular velocity $\vec{\omega}$ which makes an angle θ ($< 90°$) with the long axis of the football.

(a) What is the magnitude and direction (with respect to the long axis of the football) of the angular momentum?

(b) What is the rate of precession (neglect air friction and gravity).

(V-16) (14 points) Consider a particle moving in a central field of force. If we consider the radial motion only:

(a) What is the "effective" potential in which the radial motion occurs? (Sketch and label the contributions.)

(b) What is the condition for circular motion in terms of the "effective" potential?

(c) If the central potential is $V(r) = kr^2/2$, calculate the angular frequency for circular orbits.

(VI-1) (10 points) A rectangular thin plate of dimensions a and 2a spins about an axis along the diagonal at a constant angular velocity ω.

(a) Calculate the principal moments of inertia.

(b) What is the direction and magnitude of the angular momentum?

(c) Calculate the torques on the axis of rotation.

(VI-2) (10 points) A rocket satellite travels in a circular orbit of radius r_o. The rocket motor suddenly increases the rocket velocity by 8% in its direction of motion. What is the apogee distance of the new orbit? Sketch carefully the final orbit. Draw a sketch showing how the equivalent one-dimensional effective potential changes due to the impulse.

(VI-3) (10 points) A uniform disc of radius a and mass m rotates about a fixed axis. A massless rope is fixed to a point on the outside circumference and leads to a massless spring which is in turn fastened to a fixed point. At a radius $a/2$, another cord is fastened to a spring which connects to a mass m. Set up Lagrange's equations for the disc and the weight. (Do not solve.)

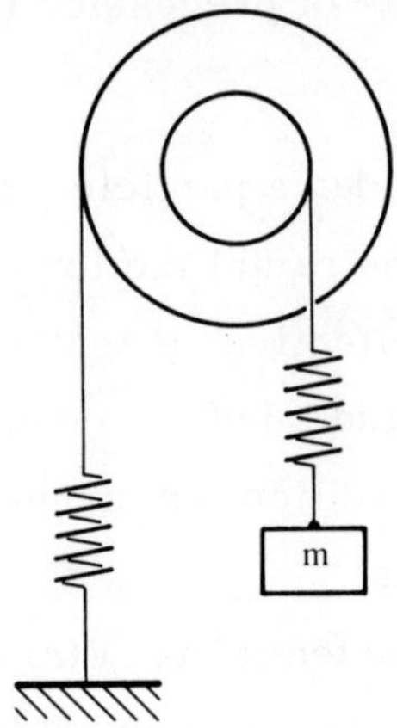

(VI-4) (10 points) A mass is attached to a spring which in turn is attached to a support. The support is driven up and down by an external mechanism such that the displacement $x = x_o \cos \omega_o t$. If there is a damping medium represented by $-b\dot{x}$, what is the amplitude of resulting motion?

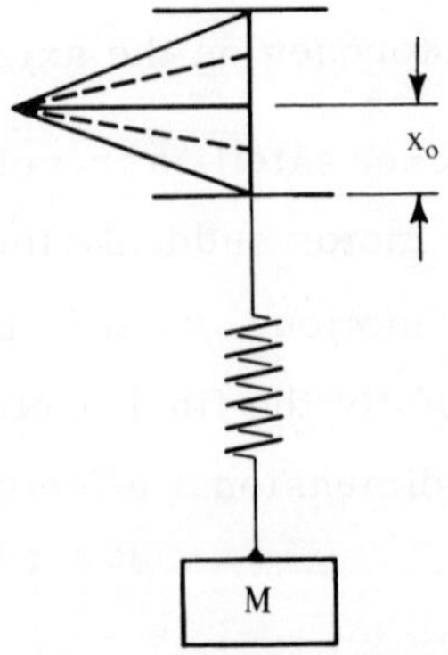

(VI-5) (10 points) A solid homogeneous cylinder of radius a rolls without slipping on the inside of a stationary larger cylinder of radius R.

(a) Write down the Lagrangian function for this system.

(b) Find the equation of motion.

(c) Find the period of small oscillations about the stable equilibrium position.

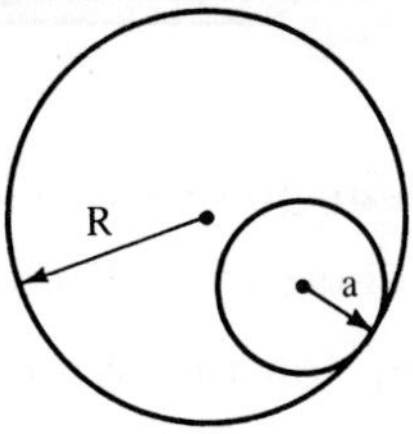

(VI-6) (10 points)

(a) Determine the radius of convergence of this power series:

$$\sum_{n=1}^{\infty} \frac{n!}{n^n} x^n$$

(b) Prove the following by the method of residues:

$$\int_{-\infty}^{\infty} \frac{\cos mx \, dx}{(x + c)^2 + a^2} = \frac{\pi}{a} e^{-a|m|} \cos cm,$$

where a is real and positive and c is real.

(VI-7) (10 points) A satellite is in a highly elliptical orbit about the earth's equator. Sketch the orbit, showing the position of the earth, and describe the effect of a small atmospheric drag upon the satellite's path. Specifically, discuss any change in the eccentricity, and the period of the orbit, the length of the major axis, and the distance of closest approach.

(VI-8) (10 points) A powerful pump is placed at A and pumps water, which flows smoothly and exits at B with a velocity of 20 cm/sec. The area at B is 6 cm^2 and the area at C is 1.0 cm^2. Find the height h of the mercury in the manometer.

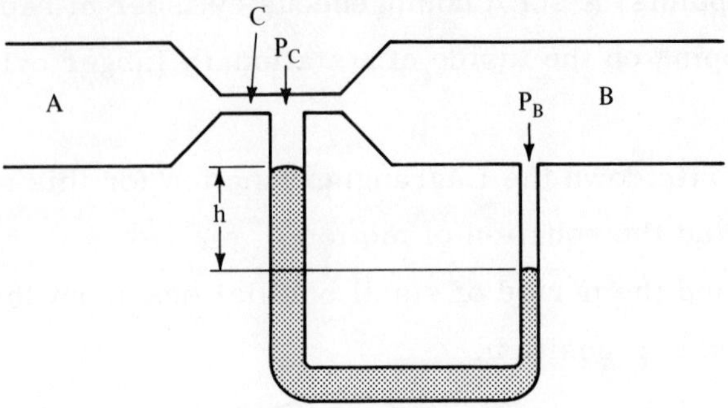

(VI-9) (5 points) Three point masses are located at the vertices of the triangle shown in the diagram. Find the principal moments of inertia about the point, P, on the hypotenuse halfway between the vertices. Indicate the orientation of the principal axes with respect to one of the sides of the triangle.

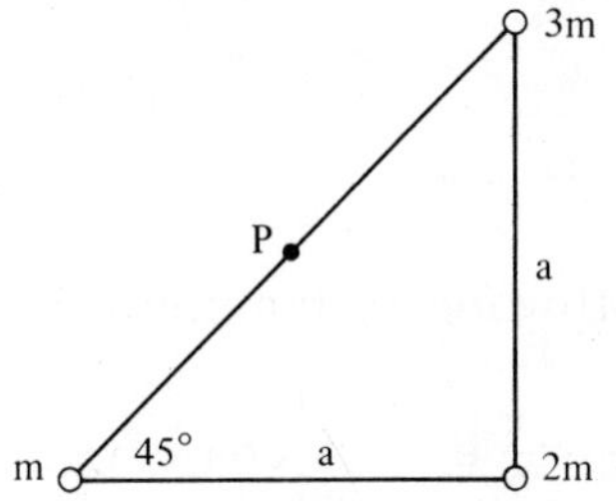

(VI-10) (10 points) A child sits in a train holding a helium-filled balloon at the end of a 4-foot nylon string. The train is traveling at 60 mph and is rounding a curve 2 miles in radius.

(a) What angle does the string of the balloon make with the vertical?

(b) Indicate the direction of the string in the horizontal plane on the sketch shown.

Answers: (a) __________

(b) 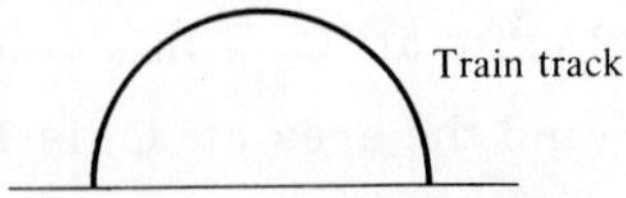

(VI-11) (5 points) The man in the figure weighs 150 lbs. He sits in a sling and pulls himself slowly up by means of a rope on a pulley. With what force must he pull? ________ lbs

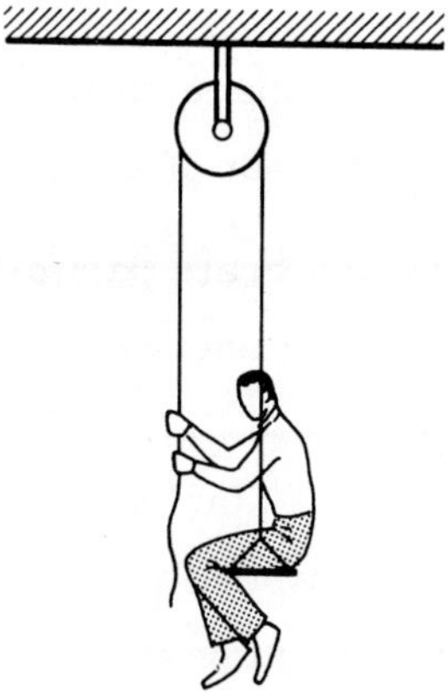

(VII-1) (20 points) Suppose you clap your hands steadily at repetition rate $1/T_1$, where T_1 is the time interval between successive claps. Each clapping sound lasts for the short time interval Δt, where Δt is short compared with T_1. During the time interval Δt we can approximate the air gauge pressure at your ear as being constant. At other times it is zero.

(a) Tell us qualitatively what will be the results of a Fourier analysis of the sound of the clapping. (You should not need actually to perform the Fourier analysis to tell us this result.)

(b) Suppose that instead of a repeated clapping sound, there is just one clap of duration Δt. Tell us qualitatively what will be the results of a Fourier analysis of this single clap. Compare the result with that in part (a).

For both parts (a) and (b), we want a sketch of the Fourier "spectrum" versus frequency, i.e., a sketch of the intensity versus frequency, with appropriate labels to show the qualitative shape of the spectrum.

(VII-2) (10 points) Given a unitary matrix S and matrix T such that $S = 1 - 2iT$:

(a) Show that in the limit as $|T_{ij}| << 1.0$, T is Hermitian.

(b) Show that for all T

$$\mathrm{Im}\,(\langle a|T|a\rangle) = -\sum_n \langle n|T|a\rangle^2$$

where a is any state in the space on which S and T operate and $|n\rangle$ represents a complete set of such states.

(VII-3) (15 points) A wooden crate partially full of bricks sits on the back of a truck. The crate is one meter square in cross section; its mass is negligible compared to that of the bricks. It is found that if the truck accelerates at a rate faster than 6 m/sec^2, the crate will begin to slide. How high may the bricks be piled into the crate before there is danger of its overturning at this acceleration?

(VII-4) (15 points) A box of delicate equipment is placed on a cushion to shield it from vertical vibration.

(a) The box sinks 6 cm. when placed on the cushion. What is the natural period of vibration of the box and cushion?

(b) If the floor vibrates at 20 cps, what is the ratio of the amplitude of vibration of the box to the amplitude of vibration of the floor?

(VII-5) (20 points) The captain of a small boat becalmed in the equatorial doldrums decides to resort to the expedient of raising the 200-kg anchor to the top of the 20-m mast. The rest of the boat has a mass of 1000 kg. (The radius of the earth is 6400 km.)

(a) The boat will begin to move. Why?

(b) Which way?

(c) How fast?

(d) Where does the energy come from?

(VII-6) (20 points) A ring with mass m_1 slides over a rod with mass m_2 and length L which is pivoted at one end and hangs vertically. Mass m_1 is secured to the pivot by a massless spring with spring constant k such that in equilibrium m_1 is centered on m_2. The motion takes place in a single vertical plane under the influence of gravity.

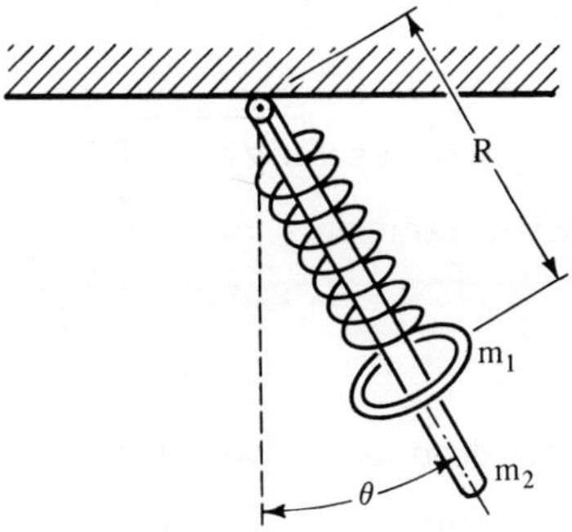

(a) Write the Lagrangian of the system as a function of θ and R as shown.

(b) Obtain the differential equations of motion for θ and R.

(c) Find one of the normal modes and its natural frequency of vibration (in the limit of small oscillations).

(VIII-1) (10 points) Show that a function that is everywhere real and analytic must be a constant.

(VIII-2) (10 points) Given a vector V and a tensor T

$$V = \begin{matrix} a_1 \\ a_2 \end{matrix}, \qquad T = \begin{matrix} a_{11} & a_{12} \\ a_{21} & a_{22} \end{matrix}$$

both referred to a two-dimensional Cartesian coordinate system x_1, x_2. Compute the components of the vector and the tensor referred to the new coordinate system obtained by a rotation of the old coordinate system by 90° in a positive sense about the $x_1 \times x_2$ axis.

(VIII-3) (20 points) Show that the function

$$f(r) = \frac{e^{-\alpha r}}{r}$$

satisfies the differential equation

$$\nabla^2 f - \alpha^2 f = -4\pi\delta(\vec{r}).$$

(VIII-4) (20 points) Find the Lagrangian and the equation of motion of the following system: m = point mass on massless rod of length L. The rod is hinged (frictionless hinge). The hinge is constrained to oscillate vertically with harmonic motion:

$$h(t) = h_o \cos \omega t.$$

The only degree of freedom is θ which measures the angle of rod from vertical. (This thing is called an "upside-down pendulum.")

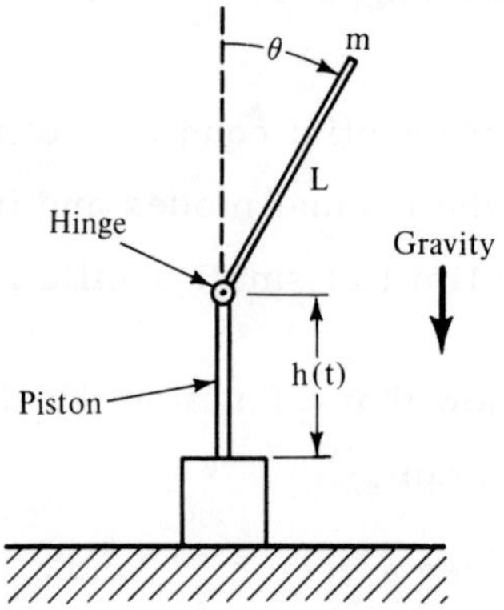

(VIII-5) (20 points) Three identical pendulum clocks are at an airport located on the earth's equator. Clock A remains at the airport. Clock E is carried by an East-flying airplane E. Clock W is carried by a West-flying airplane W. At exactly noon one day, all three clocks read the same. Planes E and W take off at noon. Plane E flies around the world at the equator, at constant ground speed, towards the East. Plane W flies around the world at the equator, at constant ground speed, towards the West. Both planes arrive back at the airport at the same time the next day. Thus both had the same ground speed. When the airplanes arrive, the clock A that remained at the airport reads exactly noon, i.e., 24 hrs, 0 min, and 0.00000000 sec have passed between take off and arrival of both planes.

(a) Do the three clocks still agree with one another? If they do not agree, list all the physical effects you can think of that would would make them not agree.

(b) Consider only the most important physical effect. Find the clock reading of clock E and that of clock W when the airport clock A reads noon (after 24-hour trip).

Assume that the radius of the earth is 6000 km.

(VIII-6) (20 points) According to the Newtonian theory of gravitation, the gravitational potential due to the sun is given (assuming spherical symmetry) by

$$-\varphi = \frac{GM}{r}$$

where $G = 67 \times 10^{-9}$ cgs units (gravitational constant), $GM = 1.3 \times 10^{26}$ cgs units, M is the mass of the sun, and r is the distance from the center of the sun.

(a) What is the formula for the force on a particle of mass m located at distance r from center of sun?

(b) According to Einstein, Newton's theory must be corrected. If we include the first-order correction, the gravitational potential can be written

$$-\varphi = \frac{GM}{r} + \frac{A}{r^2}$$

Use a dimensional argument to guess an expression for the constant A. The correction term involves the velocity of light, as well as G and M. Your guess is only expected to be good to a factor of 2.

(c) Estimate the relative magnitude of the correction term (relative to the Newtonian term) at the location of the earth, which is at $r = 1.5 \times 10^{13}$ cm.

SOLUTION SET OF MECHANICS

(I-1) Let $\vec{P}_3$, $\vec{P}_4$ be the momenta of particles 1 and 2, respectively, after collision; $|\vec{P}_4|$ is the maximum possible momentum of m_2, i.e., particle 1 is scattered backward in the c.m. system. $\vec{P}_1$, $\vec{P}_3$, $\vec{P}_4$, therefore, are in the same direction.

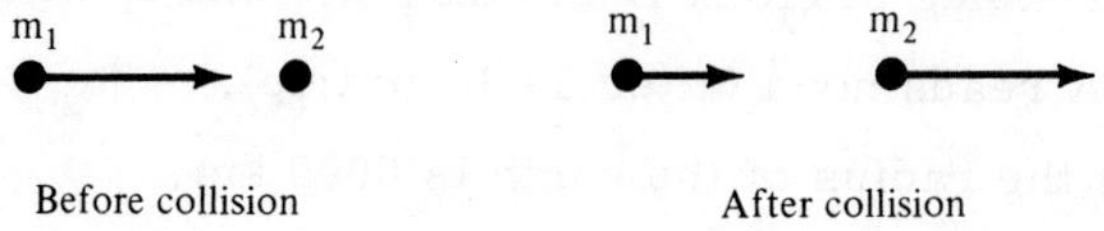

a. It is possible to solve for P_4 by the following equations according to the conservation of momentum and energy,

$$\vec{P}_1 = \vec{P}_3 + \vec{P}_4 \tag{1}$$

$$E_1 + m_2 = E_3 + E_4 \tag{2}$$

where

$$E_1 = \sqrt{\vec{P}_1^2 + m_1^2}, \quad E_3 = \sqrt{\vec{P}_3^2 + m_1^2}$$

and

$$E_4 = \sqrt{\vec{P}_4^2 + m_2^2}.$$

However it is more convenient to use the 4-vectors. Let us define:

$$P_i = (\vec{P}_i, jE_i) \quad i = 1, 2, 3, 4$$

where we have adopted the convention that

$c = 1$ and $j = \sqrt{-1}$.

(1) and (2) can be combined into the following form:

$$P_1 + P_2 = P_3 + P_4 \tag{3}$$

or

$$P_3 = P_1 + P_2 - P_4.$$

We consider the invariant scalar product

$$P_3^2 = P_1^2 + P_2^2 + P_4^2 + 2P_1 \cdot P_2 - 2P_1 \cdot P_4 - 2P_2 \cdot P_4$$

and use the relations

$$P_i^2 = -m_i^2$$

$$P_i \cdot P_j = -E_iE_j + \vec{P}_i \cdot \vec{P}_j \quad \text{for } i, j = 1, 2, 3, \text{ and } 4. \tag{4}$$

We obtain

$$-m_1^2 = -m_1^2 - 2m_2^2 - 2E_1m_2 + 2E_4m_2 + 2E_1E_4 - 2\vec{P}_1 \cdot \vec{P}_4$$

which can be simplified by the given condition that $\vec{P}_4$ is parallel to $\vec{P}_1$. We obtain

$$(m_2 + E_1)\sqrt{(P_4^2 + m_2^2)} - |\vec{P}_1| \cdot |\vec{P}_4| - (m_2^2 + E_1m_2) = 0 \tag{5}$$

or

$$|\vec{P}_4| = \frac{2m_2(m_2 + E_1)|\vec{P}_1|}{(m_2 + E_1)^2 - |\vec{P}_1|^2} \tag{6}$$

where we have discarded the other solution of (5), i.e.,

$|\vec{P}_4| = 0$, which is trivial.

b. It is given that

$$|\vec{P}_1| = m_1 = 938.2 \text{ MeV}$$

then we have

$$E_1 = \sqrt{(P_1^2 + m_1^2)} = 938.2\sqrt{2}.$$

Using $m_2 = 0.5$ MeV, we have from (6) the maximum momentum of the electron after the collision.

$$|\vec{P}_4| = \frac{2(0.5)(0.5 + 938.2\sqrt{2})}{(0.5 + 938.2\sqrt{2})^2 - (938.2)^2}(938.2)$$

$$= \frac{(1327)}{(1327)^2 - (938.2)^2}(938.2)$$

or $|\vec{P}_4| = 1.4$ MeV/c.

(I-2)

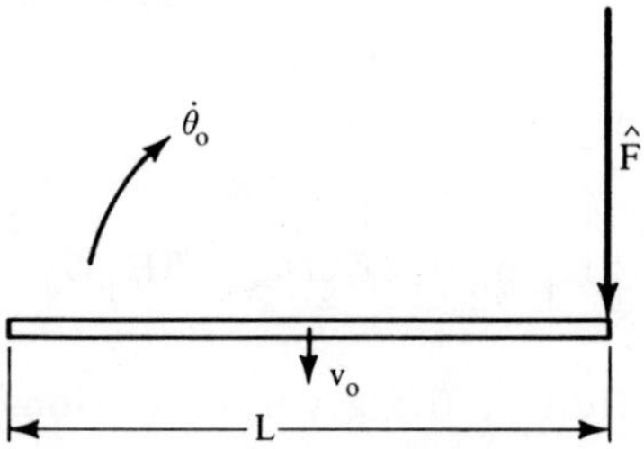

a. Let v_o be the velocity of the center of mass of the rod and $\dot{\theta}_o$ be the angular velocity of the rod with respect to its center of mass. Using the equations of impulsive motion we have

$$\hat{F} = mv_o \tag{1}$$

and

$$\frac{\hat{F}L}{2} = I\dot{\theta}_o$$

or

$$\dot{\theta}_o = \frac{\hat{F}L}{2I} \tag{2}$$

where

$$I = \frac{1}{3}m\left(\frac{L}{2}\right)^2$$

is the moment of inertia of the rod with respect to the center of mass.

If, at time t, the rod makes one complete revolution, then

$$\dot{\theta}_o t = 2\pi \tag{3}$$

or

$$t = \frac{2\pi}{\dot{\theta}_o}.$$

Therefore the distance the rod travelled after making one complete revolution is

$$S = v_o t = \frac{\hat{F}}{m} \frac{2\pi}{\dot{\theta}_o}$$

Using (2) we have

$$S = \frac{\hat{F}}{m} \frac{\pi m L^2}{3\hat{F}L} = \frac{\pi L}{3}. \tag{4}$$

b. The translational energy is

$$T_t = \frac{m}{2} v_o^2 = \frac{1}{2} m \frac{\hat{F}^2}{m^2} = \frac{1}{2} \frac{\hat{F}^2}{m}. \tag{5}$$

The rotational energy is

$$T_r = \frac{1}{2} I \dot{\theta}_o^2 = \frac{1}{2} I \left(\frac{L\hat{F}}{2I}\right)^2$$

$$= \frac{3}{2} \frac{\hat{F}^2}{m}. \tag{6}$$

The total energy is the sum of the two:

$$T(\text{total}) = T_t + T_r = \frac{2\hat{F}^2}{m}.$$

(I-3)

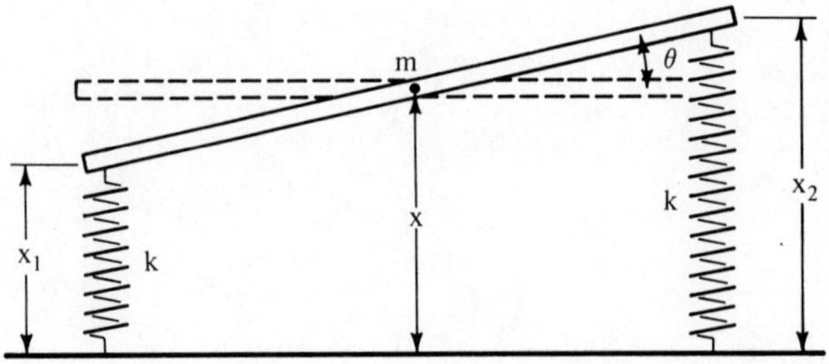

Let b be the length of spring before the motion is initiated (note that b is not the natural length of the spring because of the existence of gravitational field. b = natural length - mg/2k) and x_1, x_2, and x be the length of spring 1, spring 2, and the height of the center of mass of the rod at time t. We have the constraint that $x_1 + x_2 = 2x$ because the rod is rigid. From the second law of motion, we have

$$m\ddot{x} = -k(x_1 - b) - k(x_2 - b)$$

or

$$m\ddot{x} = -k(x_1 + x_2) + 2kb$$

$$\ddot{x} = \frac{-1}{m}(2kx - 2kb) = \frac{-2k}{m}(x - b). \tag{1}$$

Now let

$$x_2 = x + \frac{L}{2}\theta$$

and

$$x_1 = x - \frac{L}{2}\theta.$$

From the equation

$\dot{J}_o$ = torque

we get

$$I\ddot{\theta} = -k(x_2 - b)\frac{L}{2} + k(x_1 - b)\frac{L}{2}$$

or

$$I\ddot{\theta} = \frac{-kL}{2}(x_2 - x_1) = -\frac{1}{2}kL^2\theta \qquad (2)$$

where $(x_2 - x_1) = L\theta$. Using

$$I = \frac{m}{12}L^2$$

we get

$$\ddot{\theta} = \frac{-6k}{m}\theta. \qquad (3)$$

The solutions of (1) and (3) are

$$x = A\cos(\omega_1 t + B) \qquad (4)$$

and

$$\theta = C\cos(\omega_2 t + D) \qquad (5)$$

respectively, where

$$\omega_1 = \sqrt{\frac{2k}{m}} \text{ and } \omega_2 = \sqrt{\frac{6k}{m}}.$$

The initial conditions at $t = 0$ are

$$x = (b - \frac{a}{2})$$

$$\theta = \frac{a}{L}$$

$$\dot{x} = 0$$

and

$\dot{\theta} = 0.$

Substituting (6) into (5) and (4), we have

$$b - \frac{a}{2} = A \cos (B) \tag{7}$$

$$0 = -A\omega_1 \sin (B) \tag{8}$$

$$\frac{a}{L} = C \cos (D) \tag{9}$$

$$0 = -C\omega_2 \sin (D). \tag{10}$$

From (8) and (10) we get

$$B = D = 0, \tag{11}$$

$$A = b - \frac{a}{2},$$

and

$$C = \frac{a}{L}. \tag{12}$$

From (4), (5), and (12) we obtain

$$x = \frac{2b - a}{2} \cos \omega_1 t \quad \text{with} \quad \omega_1 = \sqrt{\frac{2k}{m}}$$

and

$$\theta = \frac{a}{L} \cos \omega_2 t \quad \text{with} \quad \omega_2 = \sqrt{\frac{6k}{m}}.$$

The two normal modes of vibration are defined by

$$X_1 = x_1 + x_2 = (2b - a) \cos \omega_1 t \tag{13}$$

$$X_2 = x_2 - x_1 = a \cos \omega_2 t \tag{14}$$

If $X_1 = 0$, we have $x_2 = -x_1$, i.e., mode X_2

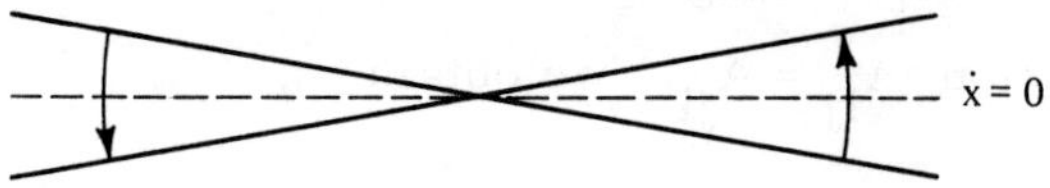

If $X_2 = 0$, we have $x_2 = x_1$, i.e., mode X_1

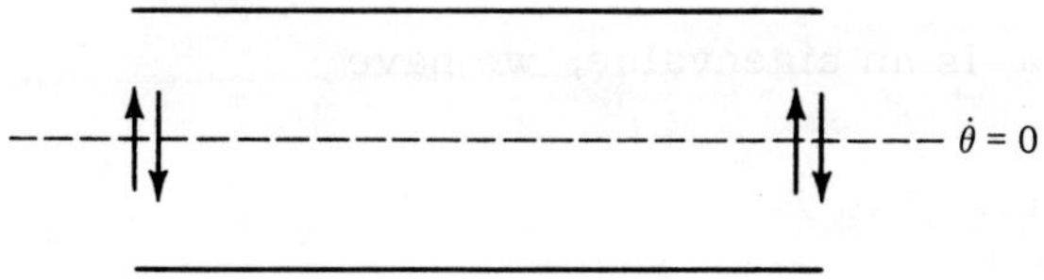

(I-4)

a. It is given that A is symmetric, i.e., $A_{ij} = A_{ji}$. We have

$$\sum_j A_{ij}\psi_j = a\psi_i \tag{1}$$

where ψ is an eigenvector of A and a is the corresponding eigenvalue of A. We multiply (1) by ψ_i^* from the left:

$$\sum_{i,j} \psi_i^* A_{ij}\psi_j = a\sum_i \psi_i^*\psi_i. \tag{2}$$

Taking the complex conjugate of (1) and multiplying on both sides by ψ_i from the left, we get

$$\sum_j A_{ij}^*\psi_j^* = a^*\psi_i^* \tag{3}$$

and

$$\sum_{i,j} \psi_i A_{ij}^*\psi_j^* = a^*\sum_i \psi_i\psi_i^*. \tag{4}$$

We can interchange the index i and j on the left side of (4) and get

$$\sum_{i,j} \psi_j A_{ij}^*\psi_i^* = a^*\sum_i \psi_i^*\psi_i. \tag{5}$$

Since A is real, $A^* = A$. Furthermore, if A is also symmetric, then $A_{ij}^* = A_{ji}$. Interchanging i and j on the left side of (5) and using the relation $A_{ij}^* = A_{ji}$, we obtain

$$\sum_{i,j} \psi_i^* A_{ij} \psi_j = a^* \sum_i \psi_i^* \psi_i . \tag{6}$$

Comparing (2) with (6), we get $a = a^*$. Therefore a is real.

b. If a is an eigenvalue, we have

$$\sum_j A_{ij} \psi_j = a \psi_i \tag{7}$$

where A_{ij} is not symmetric but real. The complex conjugate of (7) is

$$\sum_j A_{ij} \psi_j^* = a^* \psi_i^* .$$

Therefore a^* is also an eigenvalue of A.

(I-5) The equation of motion in a reference system rotating with angular velocity ω is

$$\vec{a} = \frac{\vec{F}}{m} - 2\vec{\omega} \times \vec{v} - \vec{\omega} \times (\vec{\omega} \times \vec{R}) \tag{1}$$

where

$$\frac{\vec{F}}{m} - \vec{\omega} \times (\vec{\omega} \times \vec{R}) = \vec{g}$$

is the effective acceleration of gravity and $\omega = \dfrac{2\pi}{86400} = 7 \times 10^{-5}/\text{sec}$ is the angular velocity of the earth.

(1) reduces to the following expression:

$$\vec{a} = \vec{g} - 2\vec{\omega} \times \vec{v}. \tag{2}$$

In the configuration as shown in the figure, we obtain

$$a_y = g \quad \text{or} \quad v_y = a_y t = gt \tag{3}$$

and

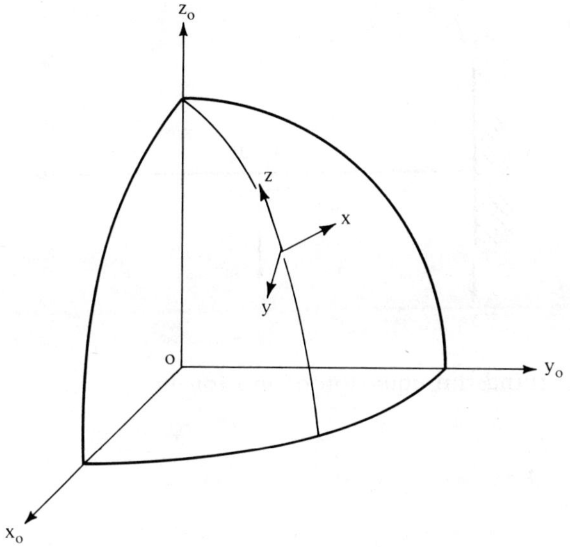

$$a_x = 2\omega v_y \sin 45° . \tag{4}$$

Therefore,

$$\dot{x} = \int_0^t a_x dt = \frac{\sqrt{2}}{2} \omega g t^2$$

and

$$x = \int_0^t \dot{x} dt = \frac{\sqrt{2}\omega g}{6} t^3 . \tag{5}$$

From (3) we have

$$h = \frac{1}{2} g t^2 \quad \text{or} \quad t = \sqrt{\frac{2h}{g}} \tag{6}$$

hence (5) becomes

$$x = \frac{2\omega}{3} \sqrt{\frac{h^3}{g}} .$$

It will land east of a plumb bob suspended from the point of release.

(I-6)

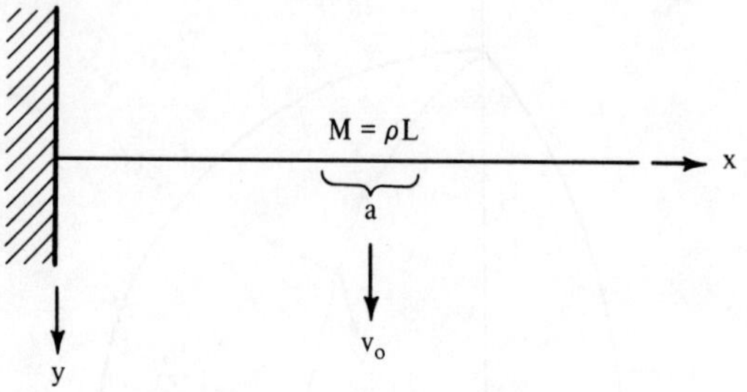

For a string the equation of motion is

$$\frac{\partial^2 y}{\partial t^2} - \frac{T}{\rho}\frac{\partial^2 y}{\partial x^2} = 0 \qquad (1)$$

where T is the tension and ρ is the mass per unit length. The initial conditions at t = 0 are given to be:

y = 0 for all x

and

$\dot{y} = 0$ for $x < (\frac{L}{2} - \frac{a}{2})$

$\dot{y} = v_o$ for $(\frac{L}{2} - \frac{a}{2}) < x < (\frac{L}{2} + \frac{a}{2})$

$\dot{y} = 0$ for $x > (\frac{L}{2} + \frac{a}{2})$.

The general solution of Equation (1) is:

$$y = \sum_{n=1}^{\infty} (A_n \cos \frac{n\pi}{L}\sqrt{\frac{T}{\rho}}\, t + B_n \sin \frac{n\pi}{L}\sqrt{\frac{T}{\rho}}\, t) \cdot \sin \frac{n\pi}{L} x \qquad (2)$$

At t = 0 we have

y = 0

That is,

$$\sum_{n=1}^{\infty} A_n \sin \frac{n\pi}{L} x = 0 \quad \text{for all } x.$$

The above relation is true only if all A_n are zero, i.e.,

$$A_n = 0 \quad \text{for all } n. \tag{3}$$

Therefore (2) leads to

$$\dot{y}(x) = \sum_{n=1}^{\infty} B_n \frac{n\pi}{L}\sqrt{\frac{T}{\rho}} \sin \frac{n\pi}{L} x. \tag{4}$$

Therefore

$$B_n = \frac{2}{n\pi}\sqrt{\frac{\rho}{T}} \int_0^L \dot{y}(x) \sin \left(\frac{n\pi x}{L}\right) dx$$

$$= \frac{2}{n\pi}\sqrt{\frac{\rho}{T}}\, v_o \int_{\frac{L-a}{2}}^{\frac{L+a}{2}} \sin \frac{n\pi}{L} x \, dx$$

$$= \frac{2v_o L}{n^2\pi^2}\sqrt{\frac{\rho}{T}} \left(\cos\left(\frac{n\pi}{2} - \frac{n\pi a}{2L}\right) - \cos\left(\frac{n\pi}{2} + \frac{n\pi a}{2L}\right)\right) \tag{5}$$

For $n = 1, 2, 3$ we can get the amplitudes of the lowest three harmonies B_1, B_2, B_3 from (5).

(I-7) In time interval Δt, the number of air molecules accelerated to the velocity V is proportional to $AV\Delta t$. Therefore the total momentum, ΔP, transferred to the air molecules is proportional to $AV^2\Delta t$. The drag force is

$$F = \frac{\Delta P}{\Delta t}$$

which is proportional to AV^2.

(I-8) In this case, the centrifugal force should equal the earth's gravitational force at the equator, i.e.,

$R\dot{\theta}^2 = g.$

Using $R = 6 \times 10^6$ m and $g = 9.8$ m/sec^2, we find

$$\dot{\theta} = \sqrt{\frac{g}{R}} \approx \frac{1}{800} \text{ sec.}$$

$$T = \frac{2\pi}{\dot{\theta}} \approx 1600\pi \text{ sec} \approx 1\tfrac{1}{2} \text{ hour.}$$

This is approximately the period of an artificial satellite around the earth at low altitude.

(I-9) Cyclic coordinates: those coordinates which do not appear explicitly in the Hamiltonian. Since

$$\dot{P}_c = \frac{\partial H}{\partial x_c} = 0,$$

the conjugate momenta of cyclic coordinates are constants of the motion.

(II-1)

a. Let a be a real eigenvalue of a skew Hermitian matrix A, i.e.

$$A\psi = a\psi. \tag{1}$$

Multiplying both sides of (1) by $\psi^\dagger$ from the left, we obtain

$$\psi^\dagger A\psi = a\psi^\dagger\psi. \tag{2}$$

The Hermitian conjugate of (1) is

$$\psi^\dagger A^\dagger = a\psi^\dagger \tag{3}$$

where we have used $a^* = a$. If we multiply (3) by ψ from the right, we obtain

$$\psi^\dagger A^\dagger \psi = a\psi^\dagger\psi. \tag{4}$$

Since $A^\dagger = -A$, (4) becomes

$$\psi^\dagger A\psi = -a\psi^\dagger\psi. \tag{5}$$

From (2) and (5) we see

$$a \equiv 0. \tag{6}$$

b. Let

$$I = \int_{-\infty}^{\infty} \frac{\sin^2 x}{x^2}\, dx$$

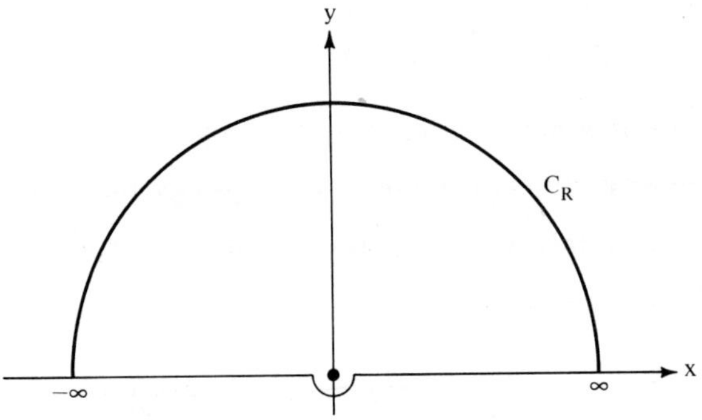

Since sin z is not well defined, when $z = x + iy$ approaches infinity, we have to express $\sin^2 x$ in terms of e^{ix}. We have

$$\sin^2 x = \frac{1 - \cos 2x}{2} = \text{Re}\left(\frac{1 - \exp(i2x)}{2}\right).$$

For positive y, exp (i2z) decreases exponentially and therefore the integral

$$\int \frac{1 - \exp(i2z)}{z^2}\, dz$$

along C_R vanishes. We can now evaluate I by substituting z for x and integrating over the upper half hemisphere.

$$I = \int_{-\infty}^{\infty} \frac{\sin^2 x}{x^2}\, dx = \text{Re}\int_{-\infty}^{\infty} \frac{1 - \exp(i2x)}{2x^2}\, dx + \text{Re}\int_{C_R} \frac{1 - \exp(i2z)}{2z^2}\, dz$$

$$= \mathrm{Re} \oint \frac{1 - \exp(i2z)}{2z^2} dz$$

$$= \mathrm{Re}\,(\pi i\ \mathrm{Res}(x = 0))$$

where Res (x = 0) is the residue at x = 0. In this case the integrand has a pole of order n = 2. The residue is

$$\mathrm{Res}\,(x = 0) = \frac{d}{dx}\left(\frac{1 - \exp(i2x)}{2}\right)\Big|_{x=0} = -i.$$

Therefore

$$I = \mathrm{Re}\,(\pi) = \pi.$$

(II-2)

a. Let us gradually peel the earth by removing one thin (outer) shell of thickness dr at a time. The energy needed to remove this shell to infinity equals the negative of the potential energy between the thin shell and the remainder. We have:

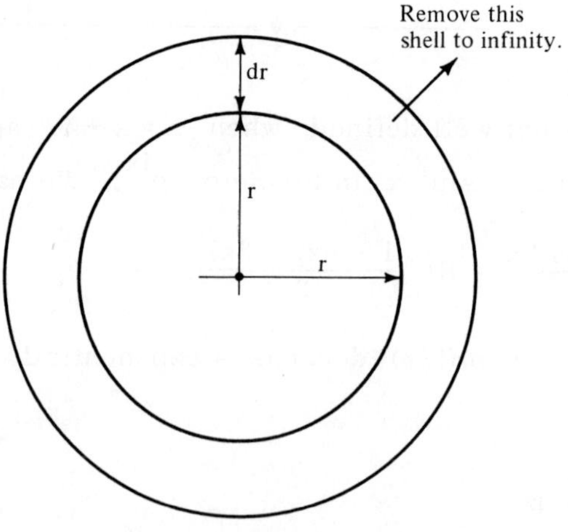

$$d(\mathrm{Work}) = -\ \text{potential energy} = G\frac{M_r m_r}{r} \tag{1}$$

where

$$M_r = \frac{4\pi}{3} r^3 \rho; \qquad m_r = -4\pi r^2 \rho\, dr$$

where ρ is the density of the earth. Here M_r is the total mass within a distance r of the center and m_r is the mass within a shell of radius r and thickness -dr. Therefore

$$d(\text{Work}) = -\,d(\text{P.E.}) = -\,G\frac{(4\pi)^2}{3}\rho^2 r^4 dr \qquad (2)$$

The total energy needed therefore is the integral of (2) from $r = R$ to $r = 0$.

$$\text{Work} = \frac{-G(4\pi)^2}{3}\rho^2\int_R^0 r^4 dr$$

$$= \frac{G(4\pi)^2}{3}\rho^2\frac{R^5}{5}$$

$$= \frac{3}{5}G\frac{M^2}{R} \qquad (3)$$

where M is the mass of the earth and R is the radius of the earth.

b. Let the potential energy $\frac{3}{5}G\frac{M^2}{R}$ equal the heat absorbed. Assuming the heat capacity of the earth is constant, we obtain the following relation:

$$\frac{3}{5}G\frac{M^2}{R} = MC_v\Delta T \qquad (4)$$

where ΔT is the increase in temperature of the earth after the earth has absorbed the amount of heat released. Assuming

$$C_v \sim 0.3\ \text{cal/gm-K°} = 1.2 \times 10^3\ \text{joule/kgm-K°},$$

we find

$$\Delta T = \frac{3}{5}G\frac{M}{R}\frac{1}{C_v}$$

$$= \frac{3}{5}gR\frac{1}{C_v}$$

$$\simeq \frac{3}{5} \times 10 \times (6 \times 10^6) \times \frac{1}{C_v} = \frac{36 \times 10^6}{C_v} \simeq 3 \times 10^4 \ K^\circ .$$

Yes, it would melt.

(II-3)

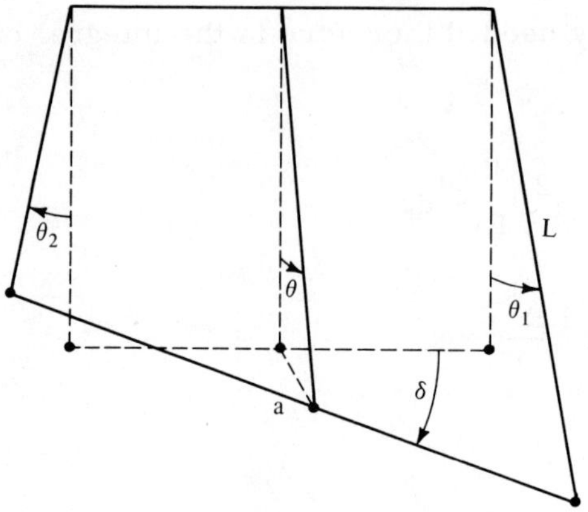

Let θ_1 and θ_2 be the angles between the vertical line and the string-one and the string-two, respectively. The angular velocity of the center of mass of the rod is

$$\dot{\theta} = \frac{1}{2}(\dot{\theta}_1 + \dot{\theta}_2).$$

The translational kinetic energy of the rod is

$$\text{T.E.} = \frac{m}{2} L^2 \dot{\theta}^2 = \frac{m}{8} L^2 (\dot{\theta}_1 + \dot{\theta}_2)^2 . \qquad (1)$$

The angle δ is related to θ_1 and θ_2 by

$$\delta = \frac{L}{a}(\theta_1 - \theta_2).$$

The rotational kinetic energy is

$$\text{R.E.} = I \frac{\dot{\delta}^2}{2} = \frac{IL^2}{2a^2}(\dot{\theta}_1 - \dot{\theta}_2)^2$$

$$= \frac{mL^2}{24}(\dot{\theta}_1 - \dot{\theta}_2)^2 \quad (2)$$

where I is the moment of inertia of the rod and equals $ma^2/12$. The potential energy (for small θ_1 and θ_2) is

$$V = \frac{1}{2}mgL((1 - \cos\theta_1) + (1 - \cos\theta_2))$$

$$\approx \frac{mgL}{4}(\theta_1^2 + \theta_2^2) \quad (3)$$

where we have made use of the series expansion for $\cos\theta$. From (1), (2), and (3) we obtain the expression for the Lagrangian $\mathcal{L}$:

$$\mathcal{L} = (T - V) = \frac{mL^2}{8}(\dot{\theta}_1 + \dot{\theta}_2)^2 + \frac{mL^2}{24}(\dot{\theta}_1 - \dot{\theta}_2)^2 - \frac{1}{4}mgL(\theta_1^2 + \theta_2^2). \quad (4)$$

where we have used T = T.E. + R.E. The two equations of motion follow from the Lagrange equation

$$\frac{mL^2}{4}(\ddot{\theta}_1 + \ddot{\theta}_2) + \frac{mL^2}{12}(\ddot{\theta}_1 - \ddot{\theta}_2) + \frac{1}{2}mgL\theta_1 = 0 \quad (5)$$

$$\frac{mL^2}{4}(\ddot{\theta}_2 + \ddot{\theta}_1) + \frac{mL^2}{12}(\ddot{\theta}_2 - \ddot{\theta}_1) + \frac{1}{2}mgL\theta_2 = 0. \quad (6)$$

We add and subtract (5) and (6) to get

$$\frac{mL^2}{3}(\ddot{\theta}_1 - \ddot{\theta}_2) + mgL(\theta_1 - \theta_2) = 0 \quad \text{or} \quad (\ddot{\theta}_1 - \ddot{\theta}_2) + \frac{3g}{L}(\dot{\theta}_1 - \dot{\theta}_2) = 0 \quad (7)$$

$$mL^2(\ddot{\theta}_1 + \ddot{\theta}_2) + mgL(\theta_1 + \theta_2) = 0 \quad \text{or} \quad (\ddot{\theta}_1 + \ddot{\theta}_2) + \frac{g}{L}(\dot{\theta}_1 + \dot{\theta}_2) = 0 \quad (8)$$

The solutions of (7) and (8) can be written in the form

$$X_1 = (\theta_1 - \theta_2) = 2A\sin\omega_1 t + 2A'\cos\omega_1 t$$

$$X_2 = (\theta_1 + \theta_2) = 2B\sin\omega_2 t + 2B'\cos\omega_2 t$$

where ω_1 and ω_2 are the corresponding frequencies. From equations (7) and (8) we have

$$\omega_1 = \sqrt{\frac{3gL}{L^2}} = \sqrt{\frac{3g}{L}} \tag{9}$$

$$\omega_2 = \sqrt{\frac{gL}{L^2}} = \sqrt{\frac{g}{L}}. \tag{10}$$

The initial conditions are

$$\theta_1 = \theta_2 = 0$$

which imply

$$A' = B' = 0. \tag{11}$$

Therefore we have

$$\theta_1 = A \sin \omega_1 t + B \sin \omega_2 t \tag{12}$$

$$\theta_2 = -A \sin \omega_1 t + B \sin \omega_2 t, \tag{13}$$

and from the impulse equations

$$\hat{P}\frac{a}{2} = I\dot{\delta} = \frac{ma^2}{12} \times \frac{L}{a}(\dot{\theta}_1 - \dot{\theta}_2) = \frac{maL}{12}(\dot{\theta}_1 - \dot{\theta}_2) \tag{14}$$

$$\hat{P} = mv = mL\dot{\theta} = m\frac{L}{2}(\dot{\theta}_1 + \dot{\theta}_2) \tag{15}$$

we can solve for $\dot{\theta}_1$ and $\dot{\theta}_2$ at $t = 0$,

$$\dot{\theta}_1 = \frac{4\hat{P}}{mL} \tag{16}$$

$$\dot{\theta}_2 = -\frac{2\hat{P}}{mL} \tag{17}$$

From (12), (13), (16), and (17) we obtain the amplitudes of the normal modes:

$$B = \frac{\hat{P}}{mL\omega_2} = \frac{\hat{P}}{m\sqrt{Lg}}$$

and

$$A = \frac{3\hat{P}}{mL\omega_1} = \frac{3\hat{P}}{m \quad 3Lg} .$$

b. 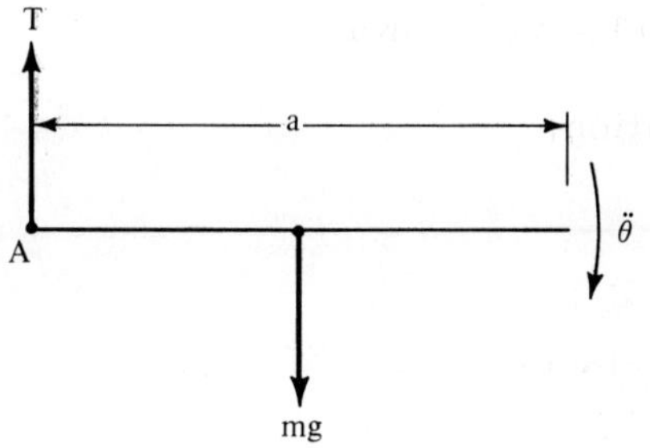

Let T be the tension of the string. According to the second law of motion we have

$$(mg - T) = m\frac{a}{2}\ddot{\theta} \tag{1}$$

and the torque equation with respect to point A is

$$mg\frac{a}{2} = I\ddot{\theta} = \frac{m}{3}a^2\ddot{\theta}. \tag{2}$$

Eliminating $\ddot{\theta}$ between (1) and (2), we get

$$T = \frac{1}{4}mg. \tag{3}$$

(II-4)

a. If the orbit is a circle, the attractive force $-dV/dr$ must equal the centrifugal force in magnitude. Therefore we have

$$\frac{v^2}{r} = 3Kr^2 \tag{1}$$

or

$$v = r\sqrt{(3Kr)} = a\sqrt{3Ka}. \tag{1}$$

The kinetic energy of the particle is

$$K.E. = \frac{1}{2}mv^2 = \frac{3Kmr^3}{2} = \frac{3Kma^3}{2} \tag{2}$$

and the angular momentum of the particle is

$$J = mr^2\dot{\theta} = mrv = mr^2\sqrt{(3Kr)} = ma^2\sqrt{3Ka} \tag{3}$$

where we have used the relation

$$\theta = \frac{v}{r} = \frac{v}{a}.$$

b. From the angular velocity

$$\dot{\theta} = \sqrt{(3Ka)},$$

we find the period of the motion to be

$$T = \frac{2\pi}{\dot{\theta}} = \frac{2\pi}{\sqrt{(3Ka)}}$$

c. After the particle is disturbed, the path changes from $r = a$ to $r = r(t)$. Let us write $r = a + x$, where x is a small quantity. The equation of motion to the first order in x is

$$m\ddot{x} + \left(\frac{3V'(a)}{a} + V''(a)\right)x = 0, \tag{4}$$

where

$$V'(a) = \left.\frac{dV(r)}{dr}\right|_{r=a} = 3Ka^2m$$

$$V''(a) = \left.\frac{d^2V}{dr^2}\right|_{r=a} = 6Kam.$$

Substituting $V'(a)$, $V''(a)$ into equation (4), we have

$$\ddot{x} + 15Kax = 0. \tag{5}$$

This is the equation of motion for a harmonic oscillator with angular frequency

$$\omega = \sqrt{15Ka}.$$

(II-5)

Before collision After collision

Let us define the following 4-vectors with the convention that $c = 1$ and $j = \sqrt{-1}$.

$$P_1 = (\vec{P}_1, jE_1), \qquad P_3 = (\vec{P}_3, jE_3),$$

$$P_2 = (\vec{P}_2, j|\vec{P}_2|), \qquad P_4 = (\vec{P}_4, j|\vec{P}_4|).$$

Using the conservation laws for momentum and energy we get

$$P_1 + P_2 = P_3 + P_4 \tag{1}$$

or

$$P_3 = P_1 + P_2 - P_4.$$

After squaring,

$$-m_e^2 = -m_e^2 + 2(E_1 - P_1 \cos\theta_{14})E_4 - 2(E_1 - P_1 \cos\theta_{12})E_2$$

$$+ 2(1 - \cos\theta_{24})E_2E_4.$$

(Note: See problem (I-1) for various details involved in this step.) We can solve for E_4:

$$E_4 = \frac{(E_1 - P_1 \cos\theta_{12})E_2}{(E_1 - P_1 \cos\theta_{14}) + (1 - \cos\theta_{24})E_2}. \tag{2}$$

For $20 = \text{GeV}$ electrons, m_e is negligible compared with E_1; therefore

$$E_1 \sim P_1.$$

To make E_4 maximum, we must have $\cos\theta_{12} = -1$,

$\cos\theta_{14} = 1$ and $\cos\theta_{24} = -1$,

so that $\vec{P}_4$ is in the same direction of $\vec{P}_1$ and opposite to $\vec{P}_2$. It follows from (2) that

$$E_4 \sim \frac{2E_1}{2E_2} E_2 \sim E_1,$$

i.e., the scattered gamma ray is almost as energetic as the incoming electron.

(III-1)

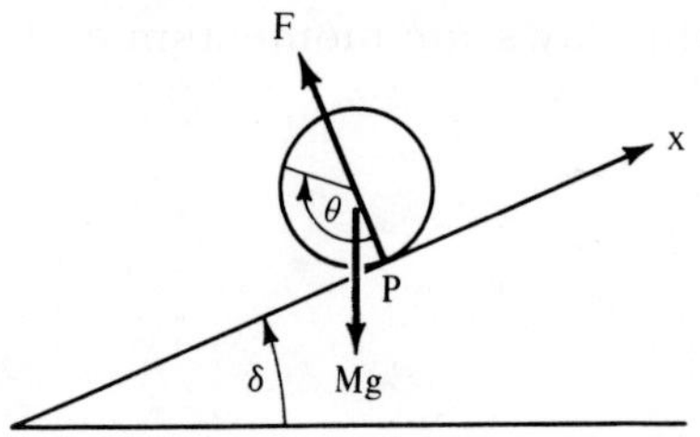

a. Let P be the point the marble is instantaneously in contact with the inclined plane at time t. The torque equation with respect to P is,

$$-\,Mg\sin\delta\, a = I_o\ddot{\theta} \tag{1}$$

where I_o is the moment of inertia with respect to point P. According to the theorem of parallel axes we have

$$I_o = \frac{2}{5}Ma^2 + Ma^2 = \frac{7}{5}Ma^2.$$

Therefore we have the linear acceleration:

$$\ddot{x} = a\ddot{\theta} = \frac{-5}{7}g\sin\delta. \tag{2}$$

b. Because of the conservation of energy, when the marble returns to its starting point, its velocity must also be v_o, but in the opposite direction. Using the relation for motion of constant acceleration

$$v_f = v_i + \ddot{x}t \tag{3}$$

where

$$v_f = -v_o, \quad v_i = v_o$$

and

$$\ddot{x} = \frac{-5}{7} g \sin \delta,$$

we obtain for (3)

$$2v_o = \frac{5}{7} gt \sin \delta . \tag{4}$$

Therefore

$$t = \frac{14v_o}{5g \sin \delta} \tag{5}$$

(III-2)

a. We are given that a particle of rest mass 1000 MeV has a momentum of 1000 MeV/c. The velocity of the particle is

$$v = \frac{Pc^2}{mc^2} = \frac{1000}{\sqrt{2} \times 1000} c = \frac{c}{\sqrt{2}} . \tag{1}$$

Its mean life is

$$\overline{T} = (1 - \beta^2)^{-1} T_o = \frac{m}{m_o} T_o = \sqrt{2} \times 10^{-8} \text{ sec.} \tag{2}$$

Therefore the mean decay distance

$$\overline{S} = v\overline{T} = 300 \text{ cm.}$$

b. Define 4-vectors:

$$P_x = (\vec{P}_x, jE_x)$$

$P_\mu = (\vec{P}_\mu, jE_\mu)$ (3)

$P_\nu = (\vec{P}_\nu, jE_\nu)$

From the conservation of energy and momentum we have

$P_\nu = P_x - P_\mu$. (5)

Squaring both sides of (5) we obtain

$P_\nu^2 = (P_x - P_\mu)^2$

or

$0 = -m_x^2 - m_\mu^2 + 2E_x E_\mu - 2P_x P_\mu \cos 15°$

Using the given values for E_x, P_μ, m_μ and m_x we obtain

$2000\sqrt{2}\, E_\mu - 2000\, P_\mu \cos 15° = 1000^2 + 100^2$. (6)

For such high energy muon the term m_μ^2 is negligible and $E_\mu \approx P_\mu$; so we get

$$E_\mu \approx \frac{500}{\sqrt{2} - \cos 15°} \approx 1100 \text{ MeV}.$$

See problem (I-1) for more details on manipulation of 4-vectors.

(III-3)

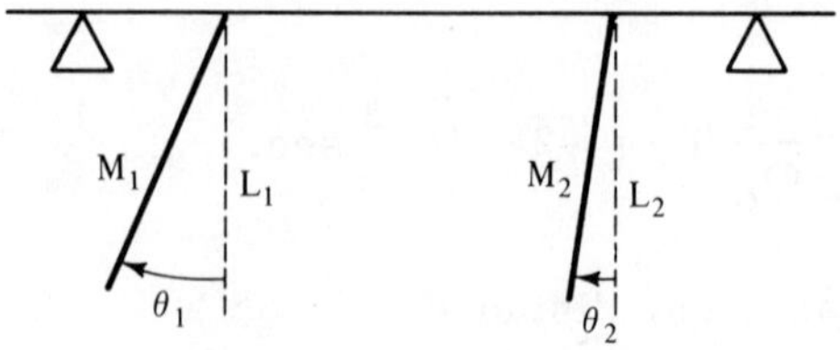

(The motion is perpendicular to the rod)

Let θ_1 and θ_2 be the angles between the vertical line and the two rods. The moments of inertia of the two rods are

$$I_1 = \frac{1}{3}M_1L_1^2 \quad \text{and} \tag{1}$$

$$I_2 = \frac{1}{3}M_2L_2^2.$$

The sum of the kinetic energies of rod one and rod two is

$$T = \frac{1}{2}I_1\dot{\theta}_1^2 + \frac{1}{2}I_2\dot{\theta}_2^2 = \frac{1}{6}M_1L_1^2\dot{\theta}_1^2 + \frac{1}{6}M_2L_2^2\dot{\theta}_2^2. \tag{2}$$

The potential energy of the whole system is

$$V = M_1g\frac{L_1}{2}(1 - \cos\theta_1) + M_2g\frac{L_2}{2}(1 - \cos\theta_2) + \frac{K}{2}(\theta_1 - \theta_2)^2 \tag{3}$$

The Lagrangian of the system is

$$L = T - V = \frac{1}{6}M_1L_1^2\dot{\theta}_1^2 + \frac{1}{6}M_2L_2^2\dot{\theta}_2^2 -$$

$$- \frac{L_1}{2}M_1g(1 - \cos\theta_1) - \frac{L_2}{2}M_2g(1 - \cos\theta_2) - \frac{K}{2}(\theta_1 - \theta_2)^2$$

$$\simeq \frac{1}{6}M_1L_1^2\dot{\theta}_1^2 + \frac{1}{6}M_2L_2^2\dot{\theta}_2^2 - \frac{1}{4}M_1L_1g\theta_1^2 - \frac{1}{4}M_2L_2g\theta_2^2 - \frac{K}{2}(\theta_1 - \theta_2)^2 \tag{4}$$

where we have made a series expansion of $\cos\theta_1$ and $\cos\theta_2$. From (4) we obtain

$$\frac{d}{dt}\left(\frac{\partial L}{\partial\dot{\theta}_1}\right) = \frac{1}{3}M_1L_1^2\ddot{\theta}_1$$

$$-\frac{\partial L}{\partial\theta_1} = \frac{1}{2}M_1L_1g\theta_1 + K(\theta_1 - \theta_2)$$

$$\frac{d}{dt}\left(\frac{\partial L}{\partial\dot{\theta}_2}\right) = \frac{1}{3}M_2L_2^2\ddot{\theta}_2$$

$$-\frac{\partial L}{\partial\theta_2} = \frac{1}{2}M_2L_2g\theta_2 - K(\theta_1 - \theta_2). \tag{5}$$

Substituting (5) into Lagrange's equation, we get the following equations of motion

$$\frac{1}{3}M_1L_1^2\ddot{\theta}_1 + \frac{1}{2}M_1L_1g\theta_1 + K(\theta_1 - \theta_2) = 0$$

$$\frac{1}{3}M_2L_2^2\ddot{\theta}_2 + \frac{1}{2}M_2L_2g\theta_2 - K(\theta_1 - \theta_2) = 0$$

or

$$2L_1\ddot{\theta}_1 + 3g\theta_1 + \frac{K}{6M_1L_1}(\theta_1 - \theta_2) = 0$$

$$2L_2\ddot{\theta}_2 + 3g\theta_2 - \frac{K}{6M_2L_2}(\theta_1 - \theta_2) = 0. \qquad (6)$$

Letting $\theta_1 = Ae^{i\omega t}$, $\theta_2 = Be^{i\omega t}$, we get for (6)

$$-2L_1\omega^2 A + 3gA + \frac{K}{6M_1L_1}(A - B) = 0$$

$$-2L_2\omega^2 B + 3gB - \frac{K}{6M_2L_2}(A - B) = 0 \qquad (7)$$

Rearranging (7) we get

$$(-2L_1\omega^2 + 3g + \frac{K}{6M_1L_1})A - \frac{K}{6M_1L_1}B = 0$$

$$-\frac{K}{6M_2L_2}A + (-2L_2\omega^2 + 3g + \frac{K}{6M_2L_2})B = 0 \qquad (8)$$

If the equations are to be consistent with each other, we must have

$$\begin{vmatrix} -2L_1\omega^2 + 3g + \frac{K}{6M_1L_1}, & -\frac{K}{6M_1L_1} \\ -\frac{K}{6M_2L_2}, & -2L_2\omega^2 + 3g + \frac{K}{6M_2L_2} \end{vmatrix} = 0 \qquad (9)$$

From equation (9), we get

$$(-2L_1\omega^2 + 3g + \frac{K}{6M_1L_1})(-2L_2\omega^2 + 3g + \frac{K}{6M_2L_2}) - \frac{K^2}{36M_1M_2L_1L_2} = 0$$

or

$$4L_1L_2\omega^4 - \left(6g(L_1 + L_2) + \frac{K}{3}\left(\frac{M_1L_1^2 + M_2L_2^2}{M_1M_2L_1L_2}\right)\right)\omega^2$$

$$+ \left(9g + \frac{2K(M_1L_1 + M_2L_2)}{4M_1M_2L_1L_2}\right)g = 0 \qquad (10)$$

From (10) we can solve for ω_1, ω_2, ω_3, and ω_4. Substituting ω_j $(j = 1, \ldots, 4)$ into (8) we can obtain various relations between A_j and B_j $(j = 1, \ldots, 4)$. The general solutions of (6) then are

$$\theta_1 = \sum_{j=1}^{4} A_j e^{i\omega_i t} \text{ and } \theta_2 = \sum_{j=1}^{4} B_j e^{i\omega_i t}.$$

(III-4) Let us define three unit vectors: $\hat{x}_1$, pointing downward to the inclined plane, $\hat{s}_1$, tangential to the track of the particle, $\hat{t}_1$ directed along the string toward the hole (see figure).

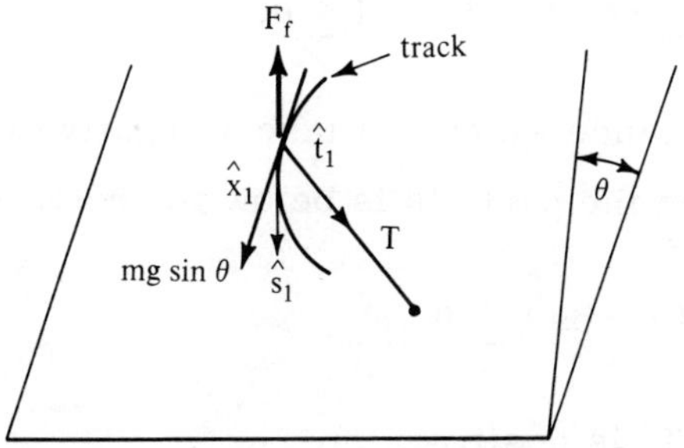

The condition of limiting equilibrium for the particle barely to move is

$$\vec{T} + mg \sin\theta\, \hat{x}_1 + \vec{F}_f = 0, \qquad (1)$$

where $\vec{F}_f$ is the friction force of the plane on the particle, which is always opposite to the direction of motion, and $\vec{T}$ is the tension in the string. By definition:

$$\vec{F}_f = -\mu F_n \hat{s}_1 = -\mu mg \cos\theta \, \hat{s}_1$$

$$= -mg \sin\theta \, \hat{s}_1$$

since $\mu = \tan\theta$. Therefore we have for equation (1)

$$\vec{T} + mg \sin\theta \, \hat{x}_1 = mg \sin\theta \, \hat{s}_1. \tag{2}$$

Squaring both sides of (2), we obtain

$$T^2 + m^2g^2 \sin^2\theta + 2Tmg \sin\theta \, (\hat{t}_1 \cdot \hat{x}_1) = m^2g^2 \sin^2\theta.$$

After simplification we get

$$T(T + 2mg \sin\theta \cos\delta) = 0 \tag{3}$$

where $\cos\delta = \hat{t}_1 \cdot \hat{x}_1$. From (3) we find when the particle is above the hole, i.e. when

$$0 \le \delta \le \frac{\pi}{2} \quad \text{and} \quad 0 \le \cos\delta \le 1, \qquad T \le 0.$$

Thus T is zero since we cannot take a negative value for the tension T. When the particle is below the hole, we find

$$\frac{\pi}{2} \le \delta \le \pi \quad \text{and} \quad -1 \le \cos\delta \le 0$$

and the tension T is positive:

$$T = -2mg \sin\theta \cos\delta > 0.$$

The path of the particle is determined by the resultant force:

$$\vec{F} = \vec{T} + mg \sin\theta \, \hat{x}_1$$

$$= mg \sin\theta \, (\hat{x}_1 - 2\cos\delta \, \hat{t}_1)$$

$$= mg \sin\theta \, (\hat{x}_1 + 2\cos\delta \, (\sin\delta \, \hat{y}_1 - \cos\delta \, \hat{x}_1))$$

$$= mg \sin\theta \, ((1 - 2\cos^2\delta)\hat{x}_1 + 2 \, c\ldots$$

$$= mg \sin\theta \, ((1 - 2\cos^2\delta)\hat{x}_1 + 2\cos\delta \sin\delta \, \hat{y}_1)$$

$$= mg \sin\theta \cos 2\delta \, (-\hat{x}_1 + \tan 2\delta \, \hat{y}_1) \tag{4}$$

See the Figure below for $\hat{y}_1$-direction. The path can be sketched as follows.

From A to B, the tension is infinitesimally small (but positive) and the particle just slides down the inclined plane. For $\delta \geq \pi/2$, from equation (4) we see the following relation must hold:

$$\frac{dy}{dx} = -\tan(2\delta) = -\frac{2\sin\delta\cos\delta}{\cos^2\delta - \sin^2\delta}$$

$$= \frac{2xy}{x^2 - y^2} \tag{5}$$

where

$x = r\cos(\pi-\delta) = -r\cos\delta$ and $y = r\sin(\pi-\delta) = r\sin\delta$.

Rearranging (5), we obtain

$$(x^2 - y^2)dy - 2xy\,dx = 0 \tag{6}$$

which has the solution

$$\frac{x^2}{y} + y = c \quad \text{or} \quad x^2 + (y - \frac{c}{2})^2 = \frac{c^2}{4} \tag{7}$$

where c is a constant determined by the condition that when $x = 0$, $c = y = BP$. Equation (7) therefore defines the path of the particle between B and P to be a half circle of radius $c/2$ and center at $x = 0$, $y = c/2$.

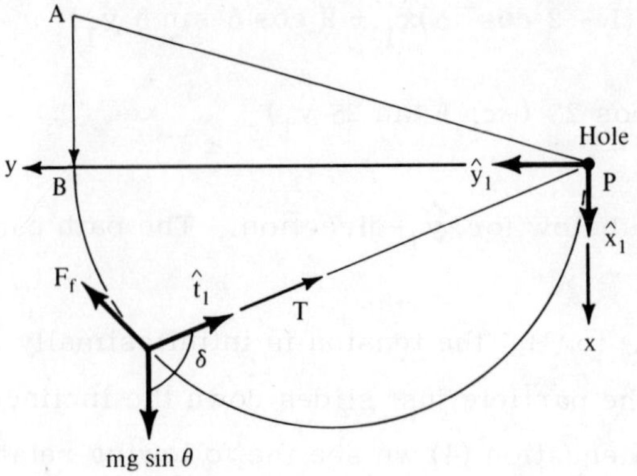

(III-5) Let P_y be the vertical force acting on the bar by the pivot. The vertical force on the pivot is $-P_y$. When the bar swings through the horizontal direction, its kinetic energy is $\frac{I}{2}\dot{\theta}^2$ which equals the difference of the initial potential energy and the final potential energy of the bar. We have the relation

$$\frac{I}{2}\dot{\theta}^2 = \frac{1}{2} LMg \sin 30°$$

where I is the moment of inertia of the bar with respect to the pivot. For a uniform rod pivoted at one end,

$$I = \frac{1}{3} ML^2 .$$

Therefore we have

$$\frac{ML^2}{6}\dot{\theta}^2 = \frac{1}{4} LMg . \tag{1}$$

Since $y = \frac{L}{2} \sin \theta$, we find

$$\dot{y} = \frac{L}{2} \cos \theta \, \dot{\theta} \quad \text{and} \quad \ddot{y} = \frac{L}{2} (\cos \theta \, \ddot{\theta} - \sin \theta \, \dot{\theta}^2).$$

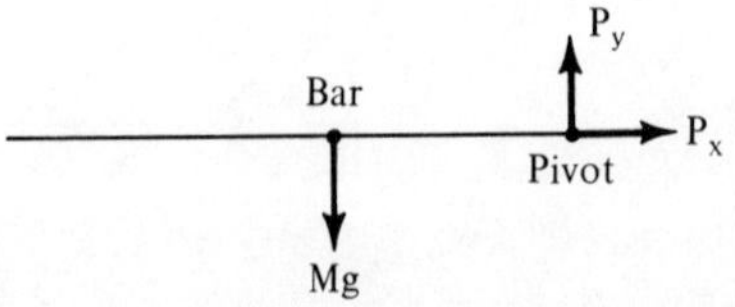

Diagram of forces on the bar.

The equation of motion is

$$Mg - P_y = M\ddot{y} = M\frac{L}{2}[\ddot{\theta} \cos\theta - (\dot{\theta})^2 \sin\theta] \tag{2}$$

When the rod is horizontal, the term $[-(\dot{\theta})^2 \sin\theta] \equiv 0$. The torque equation with respect to the center of mass of the bar is

$$P_y \frac{L}{2} \cos\theta = I_{CM}\ddot{\theta}.$$

Using

$$I_{CM} = \frac{ML^2}{12},$$

we obtain

$$P_y \cos\theta = \frac{ML}{6}\ddot{\theta}. \tag{3}$$

Eliminating $\ddot{\theta}$ between (2) and (3), we get

$$Mg - P_y = 3P_y$$

or

$$P_y = \frac{Mg}{4} \tag{4}$$

$-P_y$ is acting vertically downward on the pivot. P_x can be found from the condition that P_x has to equal the centripetal force of the bar. Therefore at the instant that the bar is horizontal

$$P_x = \int_0^L \frac{M}{L} x\dot{\theta}^2 dx$$

$$= \frac{ML\dot{\theta}^2}{2}. \tag{5}$$

Eliminating $\dot{\theta}$ between (1) and (5), we get

$$P_x = \frac{3Mg}{4}.$$

The direction of $-P_x$ is such that $-P_x$ tends to pull the pivot toward left.

(III-6) The incident energy of the moving proton at threshold is

$$T_{th} = \Delta M(1 + \frac{m_1}{m_2} + \frac{\Delta M}{2m_2}). \tag{1}$$

For the process

$$p + p \rightarrow p + p + p + \bar{p}$$

we have

$\Delta M = 2m_p$ and $m_1 = m_2 = m_p$.

Therefore

$$T_{th} = 6m_p = 5.62 \text{ BeV}.$$

For the derivation of (1), see pp. 397-400 Classical Electrodynamics by J. D. Jackson, or for a different approach see problem (XI-7) of Atomic Physics Section.

(IV-1)

a. The energy dissipated in the tide comes from the rotational energy of the earth. The maximum energy available equals the spin rotational energy of the earth which is $\frac{1}{2}I\omega^2$, where I is the moment of the inertia of the earth and ω is the spin angular velocity of the earth.

b. The earth is divided into two parts by the plane perpendicular to the earth's orbit and passing through the center of the sun and the center of the earth. Since the earth is not a perfect sphere, one part of it is slightly closer to the sun than the other. Therefore the sun exerts a greater force on it according to the inverse square law.

Let $\vec{r}_1$ and $\vec{r}_2$ be the position vectors of the center of mass of the two parts of the earth with respect to the sun; $\vec{F}_1$ and $\vec{F}_2$ be the forces on the two parts of the earth by the sun respectively and $\vec{r}_0$ be the position vector of the first part of the earth with respect to the center of mass of the earth. Therefore we have

$$\vec{r}_1 = \vec{r} + \vec{r}_0 \text{ and } \vec{r}_2 = \vec{r} - \vec{r}_0.$$

The torque on the earth by the sun is approximately

$$\vec{\tau} = 0 = \vec{r}_1 \times \vec{F}_1 + \vec{r}_2 \times \vec{F}_2 = \vec{r}_0 \times (\vec{F}_1 - \vec{F}_2) + \vec{r} \times (\vec{F}_1 + \vec{F}_2)$$

$$\equiv \vec{C} + \vec{r} \times \vec{F} \qquad (1)$$

where we have decomposed the torque into two parts: a couple $\vec{C} = \vec{r} \times (\vec{F}_1 - \vec{F}_2)$ with respect to the center of mass of the earth and a torque represented by a force $\vec{F}$ through the center of mass of the earth with respect to the sun. The couple $\vec{C}$ slows down the spin angular velocity of the earth while the torque $\vec{r} \times \vec{F}$ increases the orbital angular momentum of the earth. As L_{orbit} increases, the radius of the earth's orbit increases too, as does the period of revolution in accordance with Kepler's law ($\omega^2 r^3 = 4\pi^2 r^3 / T^2 =$ constant). From equation (1) we see that the total angular momentum is conserved. The decrease in spin angular momentum results in the increase of the same amount in the orbital angular momentum.

(IV-2) Mean life time $= \dfrac{2 \times 10^{-6}}{\sqrt{(1 - (0.8)^2}} = 3 \times 10^{-6}$ second.

(IV-3) Less.

(IV-4) Less than.

(IV-5) $\bar{x} = 1.5$ and $\bar{y} = 1$.

(IV-6) Remained the same.

(IV-7) From the vector relationship between spin $\vec{S}$, torque $\vec{L}$, and precession vector $\vec{\Omega}$

$$\vec{\Omega} \,//\, \vec{S} \times \vec{L},$$

we find that $\vec{\Omega}$ is pointing upward; therefore the top precesses counterclockwise.

(IV-8) The reduced mass of the system is

$$\frac{3 \times 1}{3 + 1} m = \frac{3}{4} m.$$

The period is

$$2\pi \sqrt{\frac{(3m)}{(4k)}} = \pi \sqrt{\frac{3m}{k}}$$

(IV-9) In static equilibrium

$$\frac{dV}{dr} = 0, \quad (\text{since } F = -\frac{dV}{dr})$$

Thus

$$-2ar^{-3} + br^{-2} = 0$$

which leads to

$$r = \frac{2a}{b}.$$

(IV-10)

a. ← → or → ←, i.e., antisymmetric motions.

b. ← ← or → →, i.e., symmetric motions.

(IV-11)

$$g = \frac{GM}{R^2}$$

or

$$\frac{\Delta g}{g} = -2\frac{\Delta R}{R}.$$

Therefore we have 2% increase in g for 1% decrease in radius of the earth.

(IV-12) Let P be the point where the can is in contact with the inclined plane at time t. The torque equation with respect to P is

$$I\ddot{\theta} = Mk^2\ddot{\theta} = MgR \sin\theta \quad \text{or} \quad k^2\ddot{\theta} = gR \sin\theta$$

where I is the moment of inertia with respect to P and k^2 is the square of the radius of gyration. We have

$$k_1^2 = \frac{I_1}{M_1} = R^2 + R^2 = 2R^2$$

for an empty can, and

$$k_2^2 = \frac{I_2}{M_2} = \frac{1}{2}R^2 + R^2 = \frac{3}{2}R^2$$

for a solid can. Since

$$k_1^2 > k_2^2,$$

we have

$$\ddot{\theta}_2 > \ddot{\theta}_1.$$

The solid can reaches the bottom first.

(IV-13)

$$\frac{\text{rate (lower hole)}}{\text{rate (upper hole)}} = \frac{\sqrt{P_\ell}}{\sqrt{P_u}} = \sqrt{3}$$

where P_ℓ(u) refers to the water pressure at lower (upper) hole.

(IV-14)

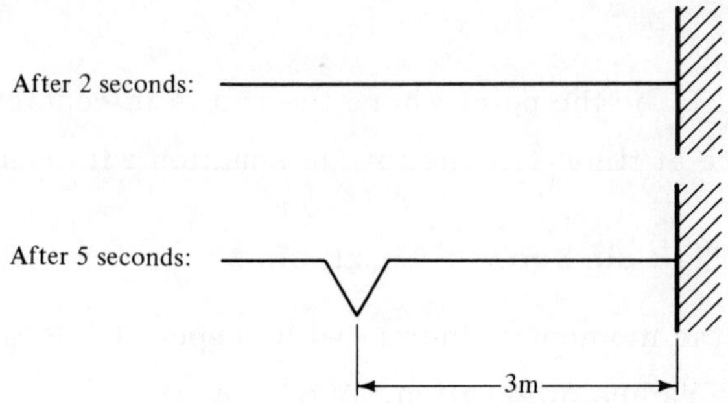

(IV-15) The Coriolis force

$\vec{F} = -2m\vec{\omega} \times \vec{v}$

a. It applies to a particle with mass m observed in a rotating coordinate system where $\vec{\omega}$ is the angular velocity of the rotating coordinate system with respect to an inertial system and $\vec{v}$ is the velocity of that particle measured in the rotating system.

b. The Coriolis force is in the west direction. Therefore the water level on the west bank is higher.

c.

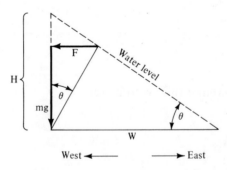

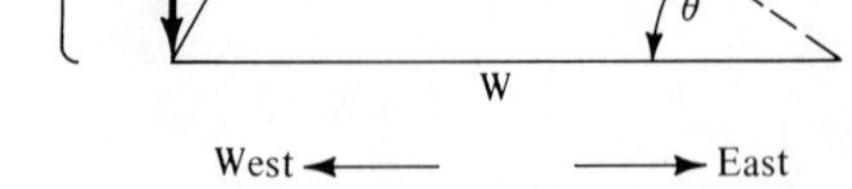

Since $|\vec{\omega} \times \vec{v}| = \omega v \sin\lambda$ we get,

$$\tan\theta = \frac{2\omega v \sin\lambda}{g} \tag{1}$$

H ≡ difference in height = W tan θ

$$= \frac{2W\omega v \sin\lambda}{g} \tag{2}$$

where v is the velocity of the water,

λ is the latitude,

ω is the earth angular velocity,

W is the width of the river,

and g is the gravitational acceleration at earth's surface.

d. $v = 5$ mi/hr $= 2.2 \times 10^2$ cm/sec

$\lambda = 30°$ or $\sin\lambda = 1/2$

$\omega = \frac{2\pi}{T} = 7.3 \times 10^{-5}$ radians/sec (where $T = 86,400$ sec)

$W = 2$ miles $\sim 3.2 \times 10^5$ cm/sec

$g \sim 10^3$ cm/sec.

Therefore (2) becomes

$$H = 2W\omega v \frac{\sin\lambda}{g}$$

$$= 2(3.2 \times 10^5)(7.3 \times 10^{-5})(2.2 \times 10^2)(\tfrac{1}{2})(10^{-3})$$

$$\sim 3.2 \times 7.3 \times 2.2 \times 10^{-1} \sim 5 \text{ cm}.$$

(IV-16)

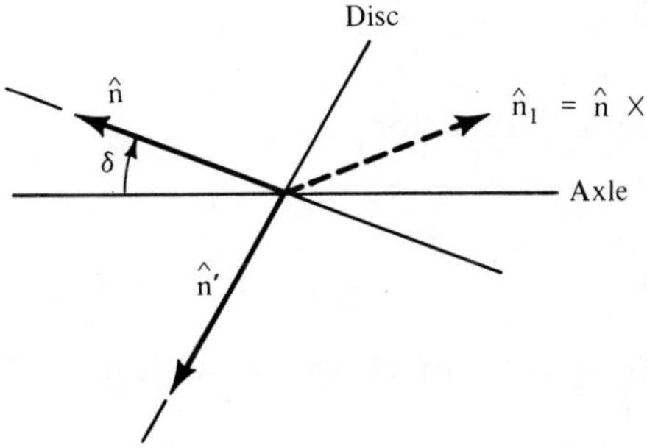

As shown in the diagram, unit vector $\hat{n}$ points in the direction of the axis of the disc; $\hat{n}'$ lies in the plane of the diagram and is perpendicular to $\hat{n}$; and $\hat{n}_1$ is perpendicular to the plane of the diagram. The angular velocity can be decomposed into two components along $\hat{n}$ and $\hat{n}'$

$$\vec{\omega} = \omega\cos\delta\,\hat{n} + \omega\sin\delta\,\hat{n}'. \tag{1}$$

The angular momentum of the disc is

$$\vec{M} = C\omega \cos\delta\ \hat{n} + A\omega \sin\delta\ \hat{n}' \tag{2}$$

The torque equation in the rotating system with angular velocity $\vec{\omega}$ is

$$\text{Torque} = \left(\frac{d\vec{M}}{dt}\right)_o = \vec{\omega} \times \vec{M} + \left(\frac{d\vec{M}}{dt}\right) \tag{3}$$

where $\left(\frac{d}{dt}\right)_o$ refers to the lab. system and $\left(\frac{d}{dt}\right)$ refers to the rotating system. The second term on the right side of (3) is zero since in this rotating frame, the disc is at rest. Therefore

$$\text{Torque} = \hat{n}_1 (A\omega^2 \cos\delta \sin\delta - C\omega^2 \sin\delta \cos\delta) = \hat{n}_1 (A-C)\omega^2 \cos\delta \sin\delta$$

where $\hat{n}_1$ is the unit vector defined by $\hat{n}_1 = \hat{n} \times \hat{n}'$.

(V-1)

a. Let x be the displacement of the block from the equilibrium position. From the equation of motion

$$M\ddot{x} + (k_2 + k_1)x = 0 \tag{1}$$

we get the frequency of vibration

$$\omega = \sqrt{\frac{(k_2 + k_1)}{M}}\,. \tag{2}$$

b. The new frequency of vibration with total mass equal to $M + m$ is

$$\omega = \sqrt{\frac{(k_2 + k_1)}{(M + m)}}\,. \tag{3}$$

c. Let v_i be the instantaneous velocity of the block when it is passing through its equilibrium position. According to the law of conservation of energy we have the relation:

$$\frac{1}{2}Mv_i^2 = \frac{1}{2}(k_1 + k_2)A^2 \qquad (4)$$

where A is the amplitude of vibration. According to the conservation of momentum, we get

$$Mv_i = (M + m)v_f. \quad (v_f = \text{velocity of the block after the mass } m \text{ is dropped on it.}) \qquad (5)$$

The conversion of kinetic energy into potential energy leads to

$$\frac{1}{2}(M + m)v_f^2 = \frac{1}{2}(k_1 + k_2)B^2. \qquad (6)$$

Using (4), (5), and (6) to eliminate v_i and v_f, we obtain the relation between the new amplitude B and A:

$$B = \sqrt{\frac{M}{(M + m)}}\,A.$$

(V-2)

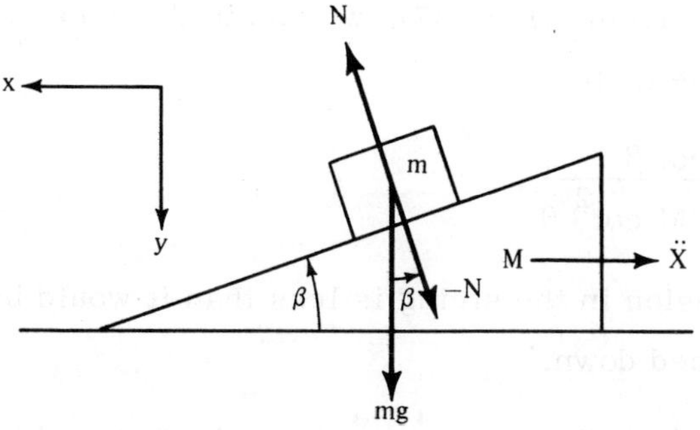

Let N be the normal force between the wedge and the mass m. For the mass m, the equation of motion in the y direction is

$$mg - N\cos\beta = m\ddot{y}. \qquad (1)$$

The equation of motion in the x direction is

$$N\sin\beta = m\ddot{x}. \qquad (2)$$

The equation of motion for the wedge in the x direction is:

$$-N \sin\beta = M\ddot{X} \qquad (3)$$

The equation defining the constraint that the mass m stays on the wedge is

$$\frac{\ddot{y}}{(\ddot{x} - \ddot{X})} = \tan\beta. \qquad (4)$$

$\ddot{y} \cos\beta = (\ddot{x} - \ddot{X}) \sin\beta$

From (2) and (3) we have

$$\ddot{x} = -\frac{M}{m}\ddot{X}. \qquad (5)$$

Eliminating N between (1) and (2) and using (5), we obtain

$$m\ddot{y} = mg + M \cot\beta\, \ddot{X} \qquad (6)$$

Eliminating $\ddot{x}$ between (4) and (5), we obtain

$$\ddot{X} = -\ddot{y} \cot\beta\left(\frac{M}{m} + 1\right)^{-1}. \qquad (7)$$

Eliminating $\ddot{y}$ from (6) and (7), we obtain the acceleration of the wedge along the table:

$$\ddot{X} = \frac{-mg \cot\beta}{(M + m) + M \cot^2\beta}.$$

(V-3) The tension in the string is less than it would be if the 5 kg block were glued down.

(V-4) The Coriolis force $-2\vec{\omega} \times \vec{v}$ is in the forward tangential direction. Therefore it is in the direction of rotation.

(V-5) Since the angular momentum has to be conserved, we have

$$J = I\dot{\theta} = \text{constant}. \qquad (1)$$

Since I is proportional to R^2, (1) becomes

$$R^2\dot{\theta} = \text{constant}$$

or

$$2\frac{\Delta R}{R} + \frac{\Delta\dot{\theta}}{\dot{\theta}} = 0.$$

Therefore, a one percent decrease in R results in a two percent increase in $\dot{\theta}$. The rotational energy is

$$R.T. = \frac{1}{2}I\dot{\theta}^2 = \frac{1}{2}J\dot{\theta},$$

which is proportional to $\dot{\theta}$. Therefore the rotational energy increases two percent.

(V-6) Let θ be the angle between the rod and the vertical line. The equation of motion is

$$I\ddot{\theta} = -(Mg \sin\theta)\frac{L}{2} - KL \sin\theta$$

For small θ,

$$\sin\theta \sim \theta$$

so

$$(\frac{1}{2}MgL + KL)\theta + I\ddot{\theta} = 0$$

or

$$(\frac{Mg}{2} + K)\theta + \frac{ML}{3}\ddot{\theta} = 0$$

where $I = \frac{1}{3}ML^2$ is the moment of inertia of the rod about one end. We obtain

$$\omega = \sqrt{\frac{(3(Mg + 2K))}{2ML}}$$

or the period

$$T = \frac{2\pi}{\omega} = 2\pi\sqrt{\frac{(2ML)}{(3(2K + Mg))}}.$$

(V-7)

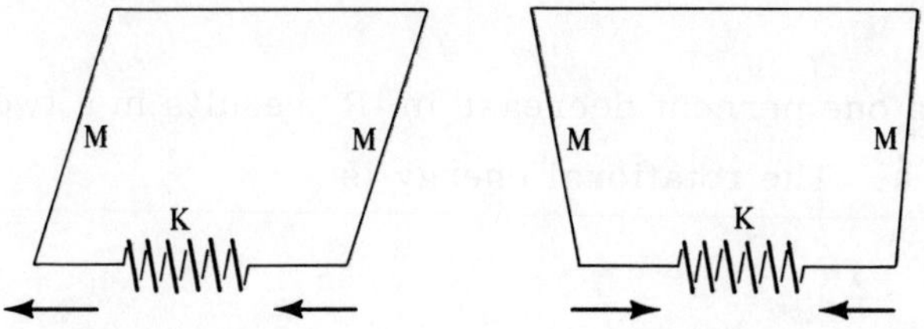

(a) Symmetric case (b) Antisymmetric case

The equations of motion for small angle motion are

$$Mg\theta \frac{L}{2} + \frac{ML^2}{3}\ddot{\theta} = 0 \quad \text{and} \quad Mg\theta \frac{L}{2} + 2KL\theta + \frac{ML^2}{3}\ddot{\theta} = 0,$$

with corresponding frequencies

$$\omega = \frac{3g}{2L} \quad \text{and} \quad \omega = \sqrt{\frac{(3(Mg + 4K))}{2ML}}.$$

(V-8) 3×10^5.

(V-9)

(c) Get there together.

(V-10)

(b) The displacement of the spring,

$d \propto (\rho_{water} - \rho_{cork}) \times \text{Volume} \times g_{effective}$. When $g_{effective}$ decreases, d decreases.

(V-11) Since the square of the period is proportional to the cube of the radius, the period of the second satellite is 1.5% larger.

(V-12)

a. At $t = 0$, $v = 0$; therefore $\frac{a_R}{a_{2R}} = 1$.

b. For two drops of the same mass, the terminal velocity is proportional to the inverse of the drag force. Therefore we get the ratio

$$\frac{v_R}{v_{2R}} = \frac{2R}{R} = 2.$$

(V-13)

a. Both forces satisfy the condition that

$$\frac{\partial F_x}{\partial y} = \frac{\partial F_y}{\partial x}, \quad \text{etc.}$$

Therefore for both forces a potential function can be defined,

$$V_1 = -2xyz + 6\, x^2 y^2 z^2 + \text{constant}$$

and

$$V_2 = -(xy^2 + yx^2 + xz^2 + zx^2 + yz^2 + zy^2 + 2xyz) + \text{constant}.$$

b. Both forces conserve the total energy.

(V-14) The equation of motion in the x direction is

$$m\ddot{x} = \mu mg$$

or

$$\ddot{x} = \mu g. \tag{1}$$

The torque equation with respect to the center of the ball is

$$I\ddot{\theta} = -\mu mgr \tag{2}$$

where I is the moment of inertia and

$$I = \frac{2}{5}mr^2.$$

Integrating (1) and (2) we obtain

$$\dot{x} = \mu gt \tag{3}$$

and

$$\dot{\theta} = -\frac{5\mu g}{2r}\,t + \omega. \tag{4}$$

The condition that at $t = t_1$ the ball begins to roll without slipping is

$$r\dot{\theta} = \dot{x}. \tag{5}$$

Using (3), (4), and (5), we obtain

$$r(\omega - \frac{5}{2r}\mu g t_1) = \mu g t_1$$

or

$$t_1 = \frac{2r\omega}{7\mu g}. \tag{6}$$

a. Using (6) and (3) we obtain the final linear velocity of the center of mass:

$$\dot{x}_{CM} = \mu g t_1 = \frac{2}{7} r\omega. \tag{7}$$

b. The distance the sphere travels before achieving this velocity is

$$S = \frac{1}{2}\mu g t_1^2 = \frac{1}{2}\mu g \left(\frac{2r\omega}{7\mu g}\right)^2$$

$$= \frac{2r^2\omega^2}{49\mu g}.$$

b. Since there is no external torque, the angular momentum with respect to the instantaneous axis must be conserved. Therefore

$$I_o\omega = I\omega_f \tag{8}$$

where I is the sphere's moment of inertia with respect to the instantaneous axis.

$$I_o = \frac{2}{5}mr^2$$

and

$$I = I_o + mr^2 = \frac{7}{5}mr^2.$$

Therefore

$$\omega_f = \frac{2\omega}{7}.$$

The linear velocity $v_f = r\omega_f = \frac{2}{7} r\omega$ which is the same as that in (7).

(V-15)

a. The angular momentum is

$$\vec{J} = I_1\omega_1\hat{i} + I_2\omega_2\hat{j} + I_3\omega_3\hat{k}$$

$$= I_1\omega \sin\theta\ \hat{i} + I_3\omega \cos\theta\ \hat{k}, \qquad (1)$$

where $\omega = \sqrt{\omega_1^2 + \omega_2^2 + \omega_3^2}$ and $\hat{k}$ is the symmetry axis. $I_1 > I_3$.

b. The precession frequency is

$$\omega_p = \frac{\omega \cos\theta\ (I_1 - I_3)}{I_1} \qquad (2)$$

which is an immediate result of Euler's dynamical equations.

(V-16)

a. The effective potential is

$$U_{eff} = \frac{J_o^2}{2mr^2} + V(r).$$

For

$$V(r) = \frac{kr^2}{2},$$

the effective potential is shown in the figure.

b. The condition for circular motion is that the total energy E is equal to the minimum value of U_{eff}. The equation

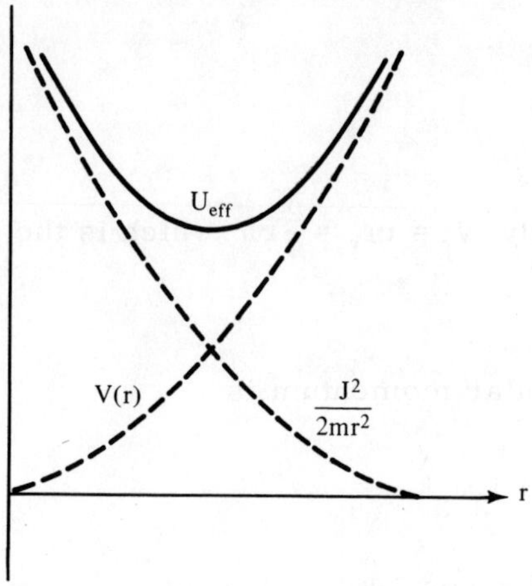

$$\left.\frac{dU_{eff}}{dr}\right|_{r=R} = 0$$

enables us to find R, the orbit radius.

c. If

$$V(r) = k\frac{r^2}{2},$$

$$U_{eff} = \frac{J_o^2}{2mr^2} + \frac{kr^2}{2}.$$

From the equilibrium condition, we have

$$-\frac{dU_{eff}}{dr} = m\frac{v^2}{R} - kR = 0$$

or

$$(\frac{v}{R})^2 = \frac{k}{m}.$$

The angular frequency is then $\sqrt{k/m}$.

(VI-1)

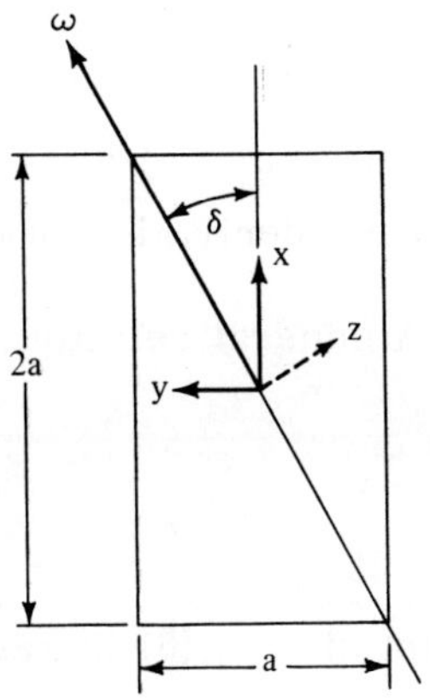

a. It is straightforward to calculate I_x and I_y:

$$I_x = \frac{1}{3}m\left(\frac{a}{2}\right)^2 = \frac{1}{12}ma^2 \tag{1}$$

$$I_y = \frac{1}{3}m\left(\frac{2a}{2}\right)^2 = \frac{1}{3}ma^2 \tag{2}$$

For a thin plate, I_z is the sum of I_x and I_y.

$$I_z = I_x + I_y = \frac{5}{12}ma^2. \tag{3}$$

b. Let $\hat{i}$, $\hat{j}$, and $\hat{k}$ be the unit vectors along the x, y, and z axes, respectively. The angular momentum with the angular velocity ω not along any of the principal axes but in the x-y plane is

$$\begin{aligned}\vec{L} &= I_x\omega_x\hat{i} + I_y\omega_y\hat{j}\\ &= I_x\omega\cos\delta\,\hat{i} + I_y\omega\sin\delta\,\hat{j}\\ &= \frac{1}{12}ma^2\omega\cos\delta\,\hat{i} + \frac{1}{3}ma^2\omega\sin\delta\,\hat{j},\end{aligned} \tag{4}$$

where we have used (1) and (2) for I_x and I_y. Note that δ is the angle between $\vec{\omega}$ and $\hat{i}$.

c. The torque on the axis of rotation is

$$\vec{N} = \left(\frac{d\vec{L}}{dt}\right)_o \tag{5}$$

where $\left(\frac{d\vec{L}}{dt}\right)_o$ means the time derivative of the vector $\vec{L}$ performed in the inertial system. A general relation is

$$\left(\frac{d\vec{L}}{dt}\right)_o = \frac{d\vec{L}}{dt} + \vec{\omega} \times \vec{L}. \tag{6}$$

The first term on the right side of (6) is zero in the rotation system; therefore the torque is

$$\vec{N} = \vec{\omega} \times \vec{L}$$

$$= \left(\frac{1}{3} ma^2 \omega^2 \sin\delta \cos\delta - \frac{1}{12} ma^2 \omega^2 \sin\delta \cos\delta\right)\hat{k}$$

$$= \frac{1}{4} ma^2 \omega^2 \sin\delta \cos\delta \; \hat{k}.$$

(VI-2) A rocket travels in a circular orbit of radius r_o. The centrifugal force must equal the inverse-square attractive force

$$mr_o \dot{\theta}^2 = \frac{k}{r_o^2}. \tag{1}$$

Therefore the angular momentum

$$J_o = m\dot{\theta} r_o^2 = \sqrt{mkr_o}. \tag{2}$$

Eight percent increase in the speed results in eight percent increase in the angular momentum too. Therefore $J = 1.08 J_o$. The differential equation for the path of the rocket is

$$\frac{d^2u}{d\theta^2} + u = \frac{km}{J^2} \tag{3}$$

where

$$u = \frac{1}{r}.$$

The solution of equation (3) is

$$r = \frac{1}{u} = \frac{1}{\frac{km}{J^2} + A\cos(\theta)} \tag{4}$$

where A is a constant. Using the initial condition

$$r = r_o = \frac{J_o^2}{km} \quad \text{for} \quad \theta = 0,$$

we obtain

$$A = \frac{1}{r_o} - \frac{km}{J^2} = \frac{1}{r_o}\left[1 - \left(\frac{J_o}{J}\right)^2\right] \simeq 0.14\frac{1}{r_o} \tag{5}$$

Therefore

$$r = \frac{r_o}{0.86 + 0.14\cos\theta} \tag{6}$$

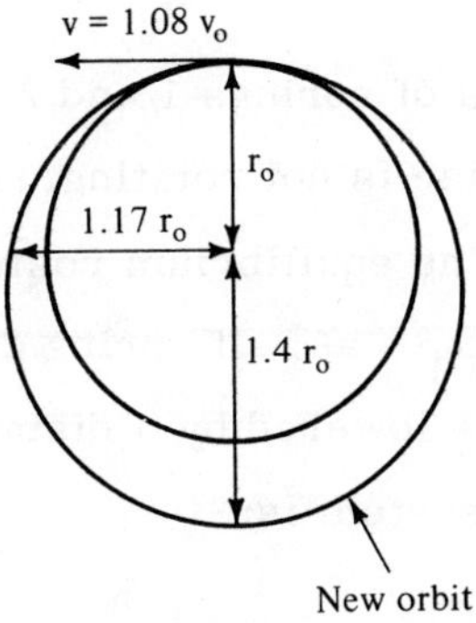

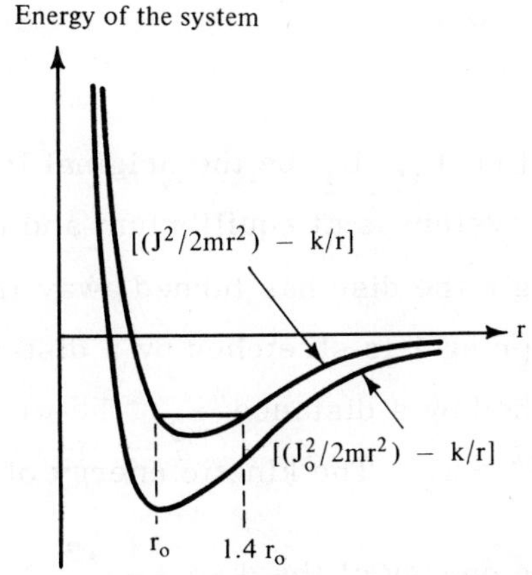

Therefore the maximum distance is $1.40r_o$.

Alternative solution:

Let $\overline{V}$ and R be the velocity and the distance of the rocket when it is at the maximum height. Since angular momentum and energy are conserved, we have

$$mvr_o = m\overline{V}R \tag{7}$$

and

$$\frac{1}{2}mv^2 - \frac{k}{r_o} = \frac{1}{2}m\overline{V}^2 - \frac{k}{R}. \tag{8}$$

From (7) and (8) we get

$$R = \frac{v^2}{-v^2 + \frac{2k}{mr_o}} r_o. \tag{9}$$

From (1) we find

$$\frac{1}{2}mv_o^2 = \frac{k}{2r_o}. \tag{10}$$

From (9) and (10) we find

$$R \approx 1.4r_o.$$

(VI-3)

Let b_1, b_2 be the original lengths of springs 1 and 2 when the whole system is at equilibrium and the disc is not rotating. If θ is the angle the disc has turned away from the equilibrium position, then spring 1 is stretched by a distance $x_1 = a\theta$. If spring 2 is stretched by a distance x_2, the weight is lowered by a distance $x_3 = \frac{a}{2}\theta + x_2$. The kinetic energy of the system is:

$$\text{Kinetic energy of the disc} = \frac{1}{2}I\dot{\theta}^2$$

$$= \frac{m}{4}a^2\dot{\theta}^2 \tag{1}$$

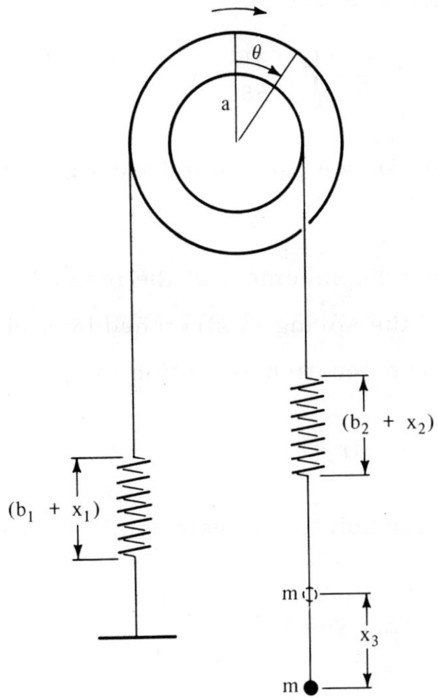

Kinetic energy of the weight $= \frac{1}{2}m\dot{x}_3^2$

$$= \frac{m}{2}\left(\frac{a}{2}\dot{\theta} + \dot{x}_2\right)^2 \qquad (2)$$

The potential energy of the system is

$$V = \frac{1}{2}ka^2\theta^2 + \frac{1}{2}kx_2^2 - mgx_3$$

$$= \frac{1}{2}ka^2\theta^2 + \frac{1}{2}kx_2^2 - mg\left(\frac{a}{2}\theta + x_2\right) \qquad (3)$$

Therefore the Lagrangian function is

$$L = T - V$$

$$= \frac{m}{4}a^2\dot{\theta}^2 + \frac{m}{2}\left(\frac{a^2}{4}\dot{\theta}^2 + a\dot{\theta}\dot{x}_2 + \dot{x}_2^2\right) - \frac{k}{2}\left(a^2\theta^2 + x_2^2\right) + mg\left(\frac{a}{2}\theta + x_2\right). \qquad (4)$$

The Lagrange's equations are

$$\frac{d}{dt}\frac{\partial L}{\partial \dot{x}_2} - \frac{\partial L}{\partial x_2} = 0 \quad \text{and} \quad \frac{d}{dt}\frac{\partial L}{\partial \dot{\theta}} - \frac{\partial L}{\partial \theta} = 0. \tag{5}$$

Substituting (4) into (5), we can immediately get two equations of motion.

(VI-4) Let x be the displacement of the mass from its equilibrium position at time t; the spring is stretched by a distance $x = x_o \exp(i\omega_0 t)$. The equation of motion is

$$m\ddot{x} + b\dot{x} + kx = kx_o \exp(i\omega_0 t)$$

which has a particular solution (steady-state term) of the form

$$x = \frac{kx_o}{\sqrt{((k - m\omega_0^2)^2 + b^2\omega_0^2)}} \cos(\omega_0 t - \delta)$$

where

$$\tan\delta = \frac{b\omega_0}{k - m\omega_0^2}.$$

The amplitude of the resulting motion after a long time is

$$\frac{kx_o}{\sqrt{((k - m\omega_0^2)^2 + b^2\omega_0^2)}}.$$

(VI-5)

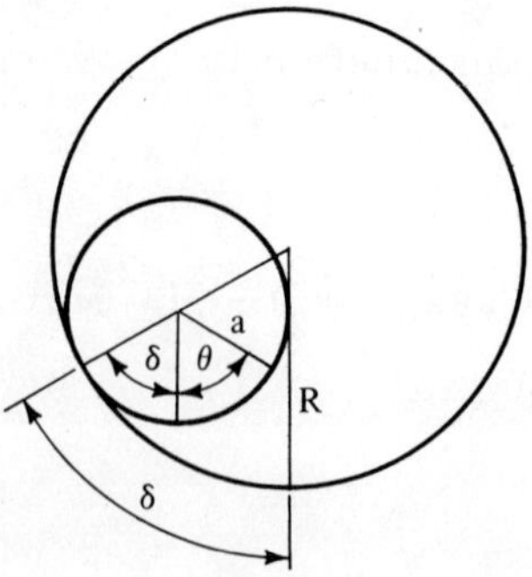

a. The kinetic energy of the cylinder can be decomposed into two parts, one involving the translation of the center of mass, and the other involving rotation about the center of mass. Let $\dot{\theta}$ be the angular velocity of the cylinder. The kinetic energy T can be written:

$$T = \frac{I_o}{2}\dot{\theta}^2 + \frac{m}{2}(R - a)^2\dot{\delta}^2 \tag{1}$$

where

$$I_o = \frac{1}{2}ma^2$$

for a solid cylinder. The constraint of rolling without sliding is $a\dot{\theta} = (R - a)\dot{\delta}$. Hence (1) becomes

$$T = \frac{m}{4}(R - a)^2\dot{\delta}^2 + \frac{m}{2}(R - a)^2\dot{\delta}^2$$

$$= \frac{3}{4}m(R - a)^2\dot{\delta}^2 \tag{2}$$

The potential energy is

$$V = -(R - a)mg\cos\delta \tag{3}$$

($V = 0$ for $\delta = 90°$)

The Lagrangian function is

$$L = T - V = \frac{3m}{4}(R - a)^2\dot{\delta}^2 + mg(R - a)\cos\delta. \tag{4}$$

b. Substituting (4) into Lagrange's equation we obtain the equation of motion

$$\frac{3}{2}(R - a)\ddot{\delta} + g\sin\delta = 0. \tag{5}$$

c. For small δ we have $\sin\delta \simeq \delta$, and (5) becomes

$$\ddot{\delta} + \frac{2g}{3(R - a)}\delta = 0$$

from which we get

$$\omega = \sqrt{\frac{2g}{(3(R - a))}}\,.$$

(VI-6)

a. To determine for which values of x the series converges, we make use of the ratio test which states that, if the absolute value of the ratio of the nth term to the (n - 1)th term in any infinite series approaches a limit $A < 1$, the series converges. We have

$$R = \frac{a_n}{a_{n-1}} = \frac{n!\,x^n(n-1)^{n-1}}{n^n(n-1)!\,x^{n-1}} = \frac{nx(n-1)^{n-1}}{n^n} = \left(\frac{n-1}{n}\right)^{n-1} x = \left(1 - \frac{1}{n}\right)^{n-1} x$$

$$\to \frac{x}{e} \text{ as } n \to \infty. \tag{1}$$

Therefore the power series converges absolutely for $|x| < e$. For $x = -e$ we find the signs of successive terms alternate and the successive terms always decrease in magnitude and the nth term approaches zero. Therefore the power series converges at $x = -e$. For $x = e$, by using Stirling asymptotic expansion:

$$\frac{n!}{n^n} = n^{1/2}e^{-n}(2\pi)^{1/2}\left[1 + \frac{1}{12n} + 0\left(\frac{1}{n^2}\right)\right].$$

We have from (1)

$$R = 1 + \frac{1}{2n} + 0\left(\frac{1}{n^2}\right).$$

Therefore the series diverges at $x = e$ according to Raabe's test.

b. The integral

$$I = \int_{-\infty}^{\infty} \frac{\cos mx\,dx}{(x + c)^2 + a^2}$$

$$= \mathrm{Re}\int_{-\infty}^{\infty} \frac{e^{imx}}{(x + c)^2 + a^2}\,dx$$

$$= \mathrm{Re} \oint \frac{e^{imz} dz}{(z + c)^2 + a^2}, \quad (z = x + iy)$$

where the integration is carried out in the upper (lower) half plane if m is greater than (less than) 0. Therefore for $m > 0$,

$$I = \mathrm{Re}\ (2\pi i\ \mathrm{Res}\ (z = -c + ia))$$

$$= \mathrm{Re}\ (2\pi i \frac{e^{im(-c + ia)}}{2ia})$$

$$= \frac{\pi}{a} \exp\ (-am) \cos cm,$$

and for $m < 0$,

$$I = \mathrm{Re}\ (\pi i\ \mathrm{Res}\ (z = -c - ia))$$

$$= \mathrm{Re}\ (2\pi i \frac{e^{im(-c - ia)}}{- 2ia}) = \frac{-\pi}{a} \exp\ (-a|m|) \cos cm.$$

(VI-7)

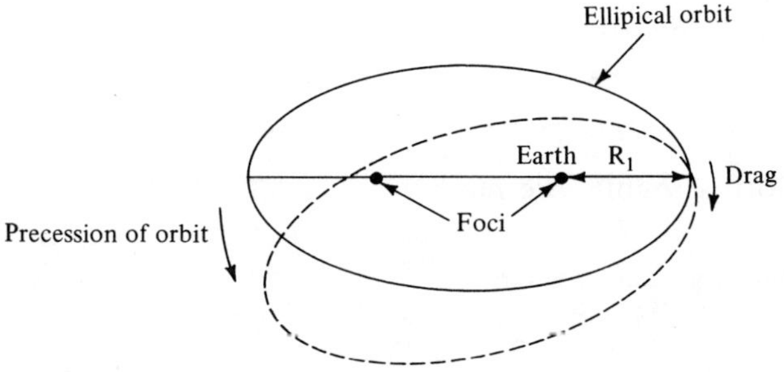

From Kepler's first law, we know the earth must be at one focus of the ellipse.

The atmospheric drag must occur when the satellite is close to the earth. Let $-\Delta P$ be the loss of linear momentum of the satellite during each revolution due to the atmospheric drag. The angular momentum loss is then

$$\Delta J = -\Delta P\ R_1 \tag{1}$$

where R_1 is the special distance from the earth to the pericenter. The eccentricity e is related to the angular momentum by

$$e = \left(\frac{2EJ^2}{mk^2} + 1\right)^{1/2} \sim 1 + \frac{EJ^2}{mk^2} \tag{2}$$

where E is the total energy of the system and k is a constant defined by potential $V = -mk/r$. For an elliptical orbit, E is negative and $0 < e < 1$. Differentiating (2), we get

$$\frac{\Delta e}{e} = \frac{J^2 \Delta E + 2EJ\Delta J}{emk^2} \tag{3}$$

where ΔE is the energy loss per revolution. Using the relations

$$\frac{J^2}{mR_1^2} = 2(E - V(R_1)) \text{ and } \Delta E = \frac{P}{m}\Delta P = \frac{J\Delta J}{mR_1^2}$$

we get

$$\frac{\Delta e}{e} = \frac{(2E - V(R_1))2J\Delta J}{emk^2}. \tag{4}$$

For an elliptical orbit, we have

$$2E - V(R_1) = mk\left(\frac{1}{R_1} - \frac{1}{a}\right) > 0 \tag{5}$$

From (4) and (5) it follows

$$\frac{\Delta e}{e} < 0, \tag{6}$$

since ΔJ is negative. From (6), we see the eccentricity is decreasing. The satellite is getting closer to the earth after each revolution. The orbit becomes more and more like a circle.

The major semiaxis, a, is proportional to $1/W$, where W is the binding energy, $W = -E$. Since the satellite is losing energy, W is increasing. Therefore a is decreasing. The period, T, is also decreasing since, according to the third law of planetary motion, the square of the period is proportional to the cube of the major semiaxis.

Since the rocket is losing energy due to the atmospheric drag when it is close to the earth, it spends more time in getting out of the atmosphere than falling into it. The angle corresponding to the closest approach is increasing for each revolution. Therefore the direction of the precession is in the same direction of the motion as shown in the figure. The distance of the closest approach R_1 as obtained from conservation of angular momentum and energy is

$$R_1 = -\frac{k}{2E} - \left(\frac{k^2}{4E^2} + \frac{J^2}{2mE}\right)^{1/2}$$

$$= -\frac{k}{2E} + \frac{k}{2E}\left(1 + \frac{2J^2E}{mk^2}\right)^{1/2}$$

$$\sim -\frac{k}{2E} + \frac{k}{2E}\left(1 + \frac{J^2E}{mk^2}\right)$$

$$\sim \frac{J^2}{2mk}$$

where we have used

$$\sqrt{E^2} = -E$$

since E is negative. Hence R_1 decreases as J decreases.

(VI-8) The pipes are horizontal; therefore, there is no potential change. The kinetic energy per unit volume is $\frac{1}{2}\rho v_C^2$ at point C and $\frac{1}{2}\rho v_B^2$ at point B. v_B is given to be 20 cm/sec. v_C can be found based on the law of conservation of material:

$$v_C = v_B \frac{\text{Area (B)}}{\text{Area (C)}}$$

$$= 120 \text{ cm/sec.}$$

The pressure difference between C and B follows from the conservation of energy for 1 cm^3 of water

$$P_B - P_C = \frac{1}{2}\rho v_C^2 - \frac{1}{2}\rho v_B^2$$

$$= 7000 \text{ gm/sec}^2/\text{cm}$$

which should equal the pressure difference due to the difference of water levels at C and B.

$$gh = P_B - P_C;$$

hence $h \sim 7$ cm.

(VI-9)

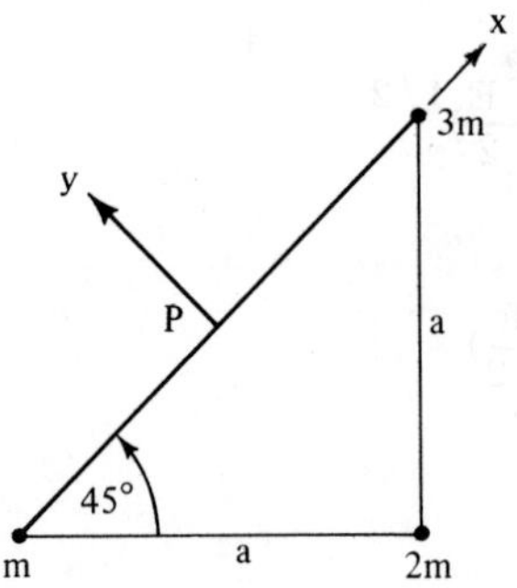

Since the three point masses are in a plane, the axis perpendicular to this plane is a principal axis at P. Let us call this the z-axis,

$$I_z = 3m\frac{a^2}{2} + 2m\frac{a^2}{2} + m\frac{a^2}{2}$$

$$= 3ma^2. \qquad (1)$$

It is easy to see the hypotenuse must be a principal axis since there is only one point mass not on this axis. Hence

$$I_x = 2m\frac{a^2}{2} = ma^2 \qquad (2)$$

Using the relation

$$I_x + I_y = I_z$$

we find I_y to be $2ma^2$.

(VI-10)

a. The density of helium is about 1/7 of the density of the air. Since the balloon is surrounded by air, one simple way to solve the problem is to imagine that the balloon is of density $-\frac{6}{7}\rho_{air}$ and neglect the effect due to the presence of the air. We find the effective gravitational acceleration is $-\frac{6}{7}g$ and the effective centrifugal acceleration is $-\frac{6a}{7}$ where

$$a = \frac{v^2}{R} = \frac{88 \times 88}{2 \times 5280} \text{ ft sec}^{-2} = 22.4 \text{ cm/sec}^2 .$$

Therefore we get

$$\tan\theta = \frac{-\frac{6}{7}a}{-\frac{6}{7}g} = \frac{a}{g} = \frac{v^2}{gR} = 0.0228$$

or

$$\theta = \tan^{-1} 0.0228 \simeq 0.023 \text{ radians} \simeq 1.32° .$$

b. In the opposite direction of the radius vector.

(VI-11) Let T be the tension of the rope. At equilibrium we have the relation

$$2T = W \quad \text{or} \quad T = 75 \text{ lbs.}$$

(VII-1) The Fourier series expansion of any function f(t) is

$$f(t) = \frac{A_o}{2} + \sum_{n=1}^{\infty} A_n \cos(nt\omega) + \sum_{n=1}^{\infty} B_n \sin(nt\omega) \qquad (1)$$

a. First we notice that the function $f(t)$ is an even function of t; therefore all B_n must vanish. Furthermore, the signal is periodic, with period T_1; so we have the condition that

$$f(t + T_1) \equiv f(t). \tag{2}$$

Using (1) and (2), we obtain

$$\omega = \omega_1 \equiv \frac{2\pi}{T_1}. \tag{3}$$

Equation (1) becomes

$$f(t) = \sum_{n=1}^{\infty} A_n \cos\left(\frac{2\pi n t}{T_1}\right) + \frac{A_o}{2}. \tag{4}$$

It is easy to see that the sum of many Fourier components with approximately equal amplitudes vanishes except in the region where most of them are in phase. In order to produce the given wave pattern, most of the Fourier components must be in phase when the wave is at its peak value, i.e.,

$$\cos(n\omega_1 t) \approx 1$$

when

$$mT_1 \le t \le (mT_1 + \Delta t) \qquad m = 1, 2, 3, \ldots$$

which leads to the condition

$$n\omega_1 \Delta t << 1 \quad \text{or} \quad n << \frac{1}{\omega_1 \Delta t}.$$

Therefore we conclude that

$$A_n \sim \text{const} \quad \text{for} \quad n << \frac{1}{\omega_1 \Delta t}$$

and the remaining components should be small enough so that they do not significantly disturb the wave pattern, i.e.,

$$A_n \to 0 \quad \text{for} \quad n > \frac{1}{\omega_1 \Delta t}$$

Therefore we get the following pattern:

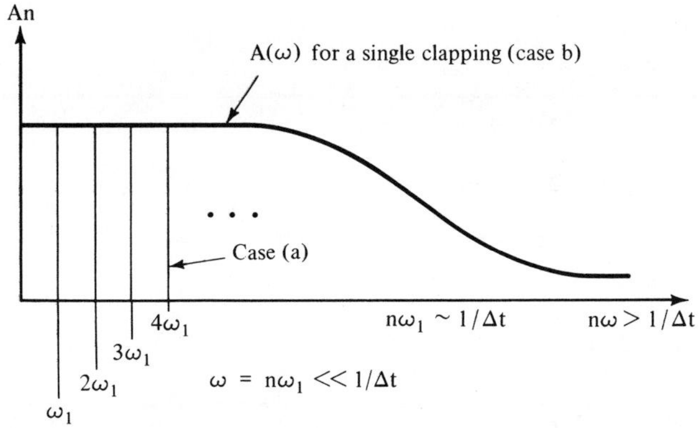

b. Let T_2 be the time interval over which we observe the single clapping. T_2 can be arbitrarily large but still finite. If we let T_1, the time interval between two successive clappings in a., be greater than T_2, we can apply the conclusion we reached in a. to the present problem. However ω_1 defined in (3) becomes very small in this case when T_2 becomes arbitrarily large. It follows that the spectrum for a single clapping is continuous.

The same conclusion is reached if we started with a Fourier transform, i.e., if we define

$$f(t) = \frac{1}{\sqrt{2\pi}} \int_{-\infty}^{\infty} e^{i\omega t} A(\omega) d\omega$$

with

$$A(\omega) = \frac{1}{\sqrt{2\pi}} \int_{-\infty}^{\infty} f(t) e^{i\omega'} dt = \frac{1}{\sqrt{2\pi}} \int_{-\Delta t/2}^{\Delta t/2} e^{i\omega' t} dt.$$

We get a continuous spectrum as shown in the figure.

(VII-2)

a. S is unitary, therefore

$$S^\dagger S = 1. \tag{1}$$

Using the given expression

$$S = 1 - 2iT$$

we obtain

$$(1 + 2iT^\dagger)(1 - 2iT) = 1$$

or

$$1 - 2i(T - T^\dagger) + 4T^\dagger T = 1. \tag{2}$$

For $|T_{ij}| << 1$, we can neglect the $T^\dagger T$ term. Equation (2) becomes

$$T - T^\dagger = 0 \text{ or } T = T^\dagger$$

i.e., T is Hermitian.

b. From equation (2) we have

$$i(T - T^\dagger) = 2T^\dagger T.$$

For the elastic case, i.e., $a \rightarrow a$, we have for (3)

$$i(< a|T|a > - < a|T^\dagger|a >) = \sum_n 2 < a|T^\dagger|n >< n|T|a >$$

where

$$\sum_n |n >< n| = 1,$$

and $|n>$ is a complete set of intermediate states. Since

$$T^\dagger_{aa} = T^*_{aa}$$

the above equation becomes

$$i(\langle a|T|a\rangle - (\langle a|T|a\rangle)^*) = \sum_n 2 \langle a|T^\dagger|n\rangle\langle n|T|a\rangle.$$

Using the relation

$$(T_{ij} - T_{ij}^*) = 2i \text{ Im } T_{ij}$$

we obtain the result

$$\text{Im } (\langle a|T|a\rangle) = -\sum_n \langle n|T|a\rangle^2.$$

(VII-3)

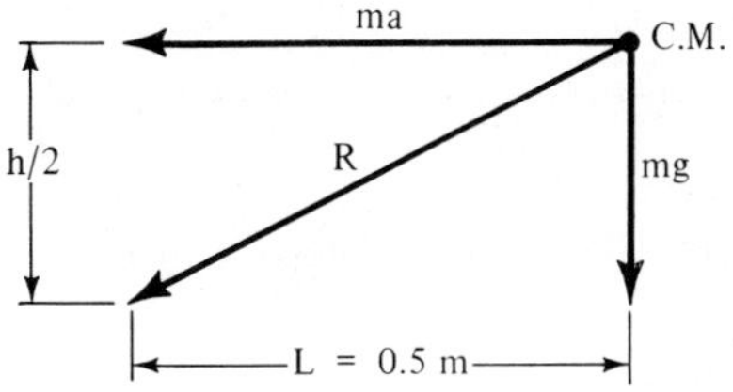

If the resultant force R points to the edge of the crate, the corresponding h will be the maximum height before the crate turns over. Therefore we have for the torque equation

$$mgL = ma\left(\frac{1}{2}h\right) \quad \text{or} \quad 9.8(0.5) = 6\left(\frac{h}{2}\right)$$

thus

$h = 1.6$ meters.

(VII-4)

a. Since the box and the cushion are in a state of equilibrium, the vertical component of the forces must vanish. Therefore

$$mg + kx = 0$$

where $x = -6$ cm is given. Therefore

$$k = \frac{mg}{6}. \tag{1}$$

The natural period of vibration is

$$T = 2\pi\sqrt{\frac{m}{k}} = 2\pi\sqrt{\frac{6}{g}} \sim \frac{1}{2} \text{ sec.}$$

b. The equation of motion when the floor is vibrating at $Ae^{i\omega_o t}$ is

$$m\ddot{x} + kx = Ake^{i\omega_o t} \tag{2}$$

where we have taken the new equilibrium position as $x = 0$. Using (1), we rewrite (2) as

$$\ddot{x} + \frac{g}{6}x = \frac{Ag}{6}e^{i\omega_o t} \tag{3}$$

from which we obtain the steady-state solution

$$x = \frac{A}{-\frac{6\omega_o^2}{g} + 1} e^{i\omega_o t}$$

Therefore the ratio of the amplitudes is

$$\frac{1}{\left|1 - \frac{6\omega_o^2}{g}\right|} \approx 0.01 \quad \text{for } \omega_o = 40\pi/\text{sec.}$$

(VII-5)

a. The vertical motion of the anchor will result in the Coriolis force

$$\vec{F} = -2m\vec{\omega} \times \vec{v}$$

acting on the boat.

b. The boat will move westward.

c. Let v be the speed of raising the anchor. Then the

acceleration of the boat is

$$a = \frac{2m\omega v}{m + M}$$

where m is the mass of the anchor and M is the mass of the boat. Integrating over the time interval T = length of the mast/v,

$$V = \int_0^T a\,dt = \frac{2m\omega}{m + M}\int_0^T v\,dt = \frac{2m\omega}{m + M}S$$

where S is the height of the mast. Therefore

$$\overline{V} = \frac{2 \times 200 \times (7.29 \times 10^{-5}) \times 20}{1200}$$

$$= 4.9 \times 10^{-4}\,\frac{\text{m}}{\text{sec.}}$$

(Note: $\omega = \frac{2\pi}{T} = \frac{6.283}{0.864 \times 10^{5}} = 7.29 \times 10^{-5}$ rad sec^{-1})

Alternative solution to (VII-5)

(c) Since angular momentum of the boat and the anchor with respect to the center of mass of the earth in an inertial frame is conserved, we have

$$(M + m)r^2\omega = (Mr^2 + m(r + s)^2)\omega'$$

or

$$\omega' \approx \omega - \frac{2ms\omega}{(M + m)r}.$$

The relative velocity with respect to the water is

$$\overline{V} = r(\omega' - \omega) = -\frac{2m\omega}{M + m}S \quad \text{(westward)}.$$

(d) The boat actually loses kinetic energy due to work done onto the centripetal force.

(VII-6)

a. Let R_o be the natural length of the spring. From the given equilibrium condition we have the following relation:

$$k\left(\frac{L}{2} - R_o\right) = m_1 g$$

Therefore

$$R_o = \frac{L}{2} - \frac{m_1 g}{k}. \tag{1}$$

The elastic potential energy of the spring stretched to the length R is

$$\frac{k}{2}(R - R_o)^2 = \frac{k}{2}\left(R - \frac{L}{2} + \frac{m_1 g}{k}\right)^2.$$

The gravitational potential energy of the rod and the ring is

$$-m_1 g R \cos\theta - m_2 g \frac{L}{2}\cos\theta.$$

(Note: The gravitational potential energy is taken to be zero when $\theta = 90°$.)

The kinetic energy is the sum of rotational energy and translational energy,

$$T = \frac{I}{2}\dot{\theta}^2 + \frac{m_1}{2}(\dot{R}^2 + R^2\dot{\theta}^2)$$

$$= \frac{1}{6}m_2 L^2\dot{\theta}^2 + \frac{m_1}{2}(\dot{R}^2 + R^2\dot{\theta}^2)$$

The Lagrangian of the system is

$$L = T - V$$

$$= \frac{1}{6}m_2 L^2\dot{\theta}^2 + \frac{m_1}{2}(\dot{R}^2 + R^2\dot{\theta}^2) + \left(m_1 g R + m_2 g\frac{L}{2}\right)\cos\theta$$

$$- \frac{k}{2}\left(R - \frac{L}{2} + \frac{m_1 g}{k}\right)^2. \tag{2}$$

b. Substituting (2) into Lagrange's equations

$$\frac{d}{dt}\frac{\partial L}{\partial \dot{\theta}} - \frac{\partial L}{\partial \theta} = 0$$

and

$$\frac{d}{dt}\frac{\partial L}{\partial \dot{R}} - \frac{\partial L}{\partial R} = 0$$

we obtain the equations of motion,

$$(m_1 R^2 + \frac{1}{3}m_2 L^2)\ddot{\theta} + (m_1 gR + m_2 g\frac{L}{2}) \sin\theta = 0 \qquad (3)$$

and

$$m_1\ddot{R} + (k - m_1\dot{\theta}^2)R + m_1 g(1 - \cos\theta) - \frac{1}{2}kL = 0 \qquad (4)$$

c. For $\theta \ll 1$ and $(R - \frac{L}{2}) \ll \frac{L}{2}$, (3) and (4) become

$$(m_1 R^2 + \frac{1}{3}m_2 L^2)\ddot{\theta} + (m_1 gR + m_2 g\frac{L}{2})\theta = 0 \qquad (5)$$

and

$$m_1\ddot{R} + k(R - \frac{L}{2}) = 0 \qquad (6)$$

where we have used $\cos\theta \approx 1$ and $\dot{\theta} = 0$. From (6) we immediately get the frequency of one of the normal modes

$$\omega = \sqrt{\frac{k}{m_1}}.$$

(VIII-1) A function that is real and analytic can be expanded into a power series

$$f(x) = \sum_{n=0}^{\infty} a_n x^n$$

but

$x^n \to \infty$ as $x \to \infty$ if $n > 0$.

Thus, if $f(x)$ is analytic, as $x \to \infty$, all $a_n (n > 0)$ must vanish. Therefore

$f(x) = a_o$ is a constant.

(VIII-2) The rotation matrix R is

$$R = \begin{pmatrix} \cos\theta & \sin\theta \\ -\sin\theta & \cos\theta \end{pmatrix}_{\theta=90^\circ} = \begin{pmatrix} 0 & 1 \\ -1 & 0 \end{pmatrix}.$$

The vector V in the new coordinate system is

$$V' = RV = \begin{pmatrix} 0 & 1 \\ -1 & 0 \end{pmatrix} \begin{pmatrix} a_1 \\ a_2 \end{pmatrix}$$

$$= \begin{pmatrix} a_2 \\ -a_1 \end{pmatrix}.$$

The tensor in the new coordinate system is

$$T' = R^\dagger TR = \begin{pmatrix} a_{22} & -a_{21} \\ -a_{12} & a_{11} \end{pmatrix}.$$

(VIII-3) If $\vec{r}$ is not zero, the left side of the differential equation $\nabla^2 f - \alpha^2 f = -4\pi\delta(\vec{r})$ can be written explicitly in spherical coordinates (r, θ, φ). We have

$$\nabla^2 f - \alpha^2 f = \frac{1}{r}\frac{\partial^2}{\partial r^2}(rf) + Lf - \alpha^2 f$$

where L is an operator depending on

$$\frac{\partial}{\partial\theta} \text{ and } \frac{\partial}{\partial\varphi}.$$

$Lf \equiv 0,$

for

$$f = \frac{e^{-\alpha r}}{r} .$$

We have

$$\nabla^2 f - \alpha^2 f = \frac{1}{r}(-\alpha)^2 e^{-\alpha r} - \alpha^2 \frac{e^{-\alpha r}}{r} = 0 \qquad (1)$$

where we have assumed that $r \neq 0$.

For $r \to 0$, we calculate the integral $I = \int (\nabla^2 f - \alpha^2 f)dV$ which integrates over a sphere of radius a with center at $r = 0$. By straightforward calculation and using the divergence theorem, we have

$$I = \int \nabla f \cdot d\vec{S} - \alpha^2 \int f dV = 4\pi - \alpha a e^{-\alpha a} - \alpha^2 \int_0^a e^{-\alpha r} 4\pi r^2 dr$$

In the limit $a \to 0$,

$$I = -4\pi \qquad (2)$$

Combining (1) and (2), we have

$\nabla^2 f - \alpha^2 f = -4\pi\delta(\vec{r})$. (Refer to (II-1) of E & M for details.)

(VIII-4) The position (x, y) of the point mass in terms of L, $h(t)$, and θ is

$$x = L \sin\theta$$
$$y = h(t) + L\cos\theta = h_o \cos\omega t + L\cos\theta . \qquad (1)$$

Differentiating (1) with respect to t we get

$$\dot{x} = \dot{\theta} L \cos\theta$$

$$\dot{y} = -(\omega h_o \sin\omega t + \dot{\theta} L \sin\theta) \qquad (2)$$

The kinetic energy is

$$T = \frac{1}{2}m(\dot{x}^2 + \dot{y}^2)$$

$$= \frac{1}{2}m(\dot{\theta}^2 L^2 \cos^2\theta + \dot{\theta}^2 L^2 \sin^2\theta + \omega^2 h_o^2 \sin^2 \omega t + 2\omega h_o \dot{\theta} L \sin\theta \sin\omega t)$$

$$= \frac{1}{2}m(\dot{\theta}^2 L^2 + \omega^2 h_o^2 \sin^2 \omega t + 2\omega h_o \dot{\theta} L \sin\theta \sin\omega t). \quad (3)$$

The potential energy is

$$V = mgy$$

$$= mg(h_o \cos\omega t + L\cos\theta). \quad (4)$$

The Lagrangian is

$$L = T - V = \frac{m}{2}(\dot{\theta}^2 L^2 + \omega^2 h_o^2 \sin^2 \omega t + 2\omega h_o \dot{\theta} L \sin\theta \sin\omega t$$

$$- 2g(h_o \cos\omega t + L\cos\theta)). \quad (5)$$

Substituting (5) into the Lagrange's equation

$$\frac{d}{dt}\frac{\partial L}{\partial\dot{\theta}} - \frac{\partial L}{\partial\theta} = 0$$

we get the equation of motion

$$L^2\ddot{\theta} + \omega^2 h_o L \sin\theta \cos\omega t + \omega h_o L\dot{\theta} \cos\theta \sin\omega t$$

$$- \omega h_o L\dot{\theta} \cos\theta \sin\omega t - gL \sin\theta = 0$$

or

$$L\ddot{\theta} + \omega^2 h_o \sin\theta \cos\omega t - g\sin\theta = 0.$$

In the case of small oscillation, we redefine $\theta' = \theta - \pi$; thus

$\sin\theta = -\sin\theta' = -\theta'$.

We have then

$$L\ddot{\theta}' + (g - \omega^2 h_o \cos\omega t)\theta' = 0.$$

If the hinge is fixed, i.e., $h_o = 0$, this is the equation of motion for a simple pendulum.

(VIII-5)

a. No, they are not the same because of the presence of the following effects:

i. Coriolis acceleration: $-2\vec{\omega} \times \vec{v}$.

ii. Gravitational force decreases as distance increases.

iii. Centripetal acceleration:

$$\frac{v^2}{r}\hat{r}_1 .$$

iv. Friction forces between the air and the pendulum.

v. Relativistic effect:

$$t = \left(1 - \frac{v^2}{c^2}\right)^{-1/2} \tau .$$

vi. Acceleration at taking-off and landing.

vii. Acceleration due to the orbital movement of the earth, etc.

b. Let us use the inertial system with the origin of the coordinates sitting at the center of the earth as the reference system. The Coriolis force $(2m\vec{\omega} \times \vec{v})$ vanishes in this system. The effective acceleration of gravity $\vec{g}$ at the earth's surface for an object with angular velocity ω is

$$\vec{g}_{eff} = -\frac{GM}{R^2}\hat{R} - \vec{\omega} \times (\vec{\omega} \times \vec{R})$$

$$= \left(-\frac{GM}{R^2} + \omega^2 R\right)\hat{R} \equiv g\hat{R} \quad (1)$$

i.e.,

$$g_{eff} \equiv g = \left(\frac{GM}{R^2} - \omega^2 R\right) = g_o - \omega^2 R$$

where $g_o = GM/R^2$. The period of a simple pendulum is known to be

$$T = 2\pi\sqrt{\frac{L}{g}}$$

where L is the constant length of the pendulum. The jet W flies westward. The angular velocity of the plane W observed in the inertial system is zero. Therefore

$$T_W = 2\pi\sqrt{\frac{L}{g_o}} = 2\pi\sqrt{\frac{L}{g + \omega^2 R}} = 2\pi\sqrt{\frac{L}{g}}\sqrt{\frac{1}{1 + \frac{\omega^2 R}{g}}} \approx T_o\left[1 + \frac{\omega^2 R}{g}\right]^{-1/2} \quad (2)$$

where $T_o = 2\pi\sqrt{\frac{L}{g}}$ is the period of the clock sitting at the airport. The jet E flies eastward. The angular velocity of the jet is 2ω, where ω is the angular velocity of the earth. Therefore

$$T_E = 2\pi\sqrt{\frac{L}{(g_o - 4\omega^2 R)}} = 2\pi\sqrt{\frac{L}{g - 3\omega^2 R}} = 2\pi\sqrt{\frac{L}{g}}\sqrt{\frac{1}{1 - \frac{3\omega^2 R}{g}}}$$

$$= T_o\left[1 - \frac{3\omega^2 R}{g}\right]^{-1/2} \quad (3)$$

Using $\omega^2 R << g$, the periods of the two pendulums can be approximated by

$$T_E = T_o + \frac{1}{2}T_o\frac{3\omega^2 R}{g_o - \omega^2 R} \equiv T_o + \frac{3\Delta T}{2}$$

and

$$T_W = T_o - \frac{1}{2}T_o\frac{\omega^2 R}{g_o - \omega^2 R} \equiv T_o - \frac{\Delta T}{2}$$

Taking

$$g = g_o - \omega^2 R = 9.8\frac{m}{sec^2}$$

$R = 6 \times 10^6 m$

$T_o = 24$ hrs

and

$$\omega = \frac{2\pi}{T_o},$$

we obtain

$$\Delta T = \frac{4\pi^2 \times 6 \times 10^6}{9.8 \times 3600 \times 24}$$

$$= 280 \text{ sec.}$$

Since T_E is larger, the clock E goes slower while the clock W goes faster. When the two jets arrive at the airport, clock E is slower by 420 seconds while clock W is faster by 140 seconds. Both are compared with the clock at the airport.

(VIII-6)

a. Since the sun is a sphere, the force is simply

$$f = \frac{-GmM}{r^2} . \tag{1}$$

b. According to general relativity, the gravitational potential is of the form

$$-\varphi = \frac{GM}{r} + \frac{A}{r^2} \tag{2}$$

where A can only depend on c, G, and M.

Let D(A) stand for the dimension of A. From (2) we find the following relation,

$$D(A) = D(GMr). \tag{3}$$

Using the relation

$$D(\frac{GmM}{r}) = D(mc^2) \equiv D(\text{energy})$$

we obtain

$$D(r) = D(\frac{GM}{c^2}). \tag{4}$$

From (3) and (4) we have

$$D(A) = D(\frac{G^2M^2}{c^2}).$$

Therefore

$$A \sim \frac{G^2M^2}{c^2}.$$

c. The ratio of the second term to the first term in (2) is

$$\frac{GM}{rc^2} = \frac{1.3 \times 10^{26}}{1.5 \times 10^{13} \times 9 \times 10^{20}} \approx 10^{-8}$$

which is very small for the earth. Mercury is the only one which shows a large effect.

ELECTRICITY AND MAGNETISM

(I-1) (20 points)

(a) What is the ratio of the skin depth in copper at 1Kc/sec to that at 100 Mc/sec?

(b) What is the electric field associated with a laser beam having an energy density 10^6 joules/cm^3?

(c) What is the relation between R, L, and C for critical damping in a series LRC circuit?

(I-2) (20 points) Suppose an electron is oscillating in a SHO potential with an angular frequency $\omega \approx 10^{15}$ rad/sec and amplitudes $A = 10^{-8}$ cm.

(a) Calculate the amount of energy radiated per cycle.

(b) What is ratio of the radiated energy per cycle to the mechanical average energy?

(c) How long will it take the system to radiate away half of its energy?

(I-3) (20 points)

(a) What is the interaction energy between two parallel dipoles, separated by a distance d? Assume that the vector joining them is perpendicular to the direction of the dipole moments.

(b) Two conducting spheres, each of radius R, are placed at a distance d from each other. There is a uniform electric field, $\vec{E}$, perpendicular to the line joining them. Assuming $R \ll d$ find the force between the spheres.

(I-4) (20 points) Consider a cylinder of radius a and length L filled uniformly with a completely ionized gas of charge density ρ which is moving parallel to the axis of the cylinder with a velocity v.

(a) Find the magnetic field at a distance r from the axis. (Neglect end effects.)

(b) Suppose a parallel beam of energetic protons of mass m, velocity V' are shot into this cylinder with their initial velocity parallel to the axis. This system can be used to focus the protons to a point on the axis. Assuming that L is much smaller than the focal length, and neglecting electrostatic and relativistic effects, calculate the focal length. (Focal length ≡ distance from the end of the cylinder to the focus.)

(I-5) (20 points) Starting from the fact that (A_x, A_y, A_z, iV) is a 4-vector, calculate the field of a point charge in uniform motion by making a Lorentz transformation from the rest frame of the charge to the laboratory.

(II-1) (15 points) Prove that $\nabla^2 \frac{1}{|\vec{r}|} = -4\pi\delta(\vec{r})$.

(II-2) (15 points) Four positive and four negative charges are alternated at eight corners of a cube such that the three charges adjacent to each charge are opposite in sign to that charge. What is the radial dependence of the magnitude of the resulting electrostatic field at large distances?

(II-3) (15 points) Two electric dipoles lying on the x-axis and oriented along the z-axis oscillate exactly out of phase. Their x-coordinates are separated by $\lambda/2$. Calculate the Poynting vector at large distances.

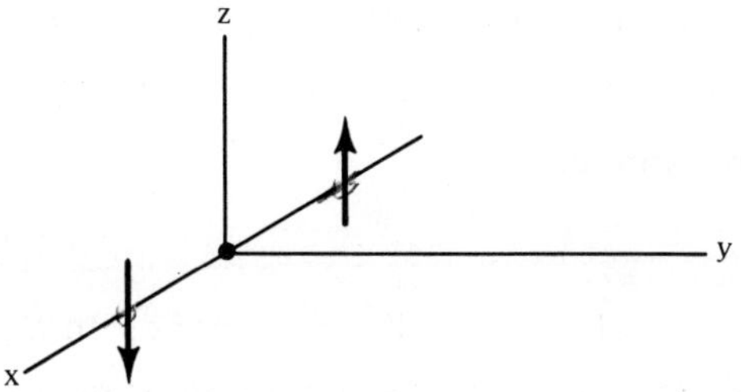

(II-4) (15 points) Two identical and coaxial superconducting loops each of self-inductance L are far apart. Each has a current I flowing in the same direction. They are then superposed. What is the final current $I'_{1,2}$ in each loop? What are the initial and final energies of the system? Account for any energy changes.

(II-5) (15 points) In the following circuit, patented by Steinmetz, the applied voltage is at a frequency of $\omega = 1\sqrt{LC}$. Determine the amplitude and phase of the current through the resistor in terms of the voltage and the circuit parameters.

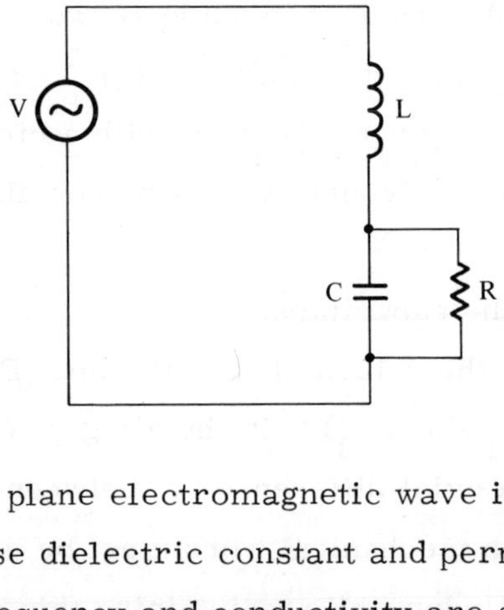

(II-6) (15 points) A plane electromagnetic wave is incident normally on a conductor whose dielectric constant and permeability are that of free space. The frequency and conductivity are such that, within the conductor, the conduction current and displacement current are equal in magnitude. What is the reflection coefficient, i.e., the ratio of reflected energy to incident energy?

(II-7) (10 points) An uncharged conducting sphere is placed in a uniform electric field. What is the angular and radial dependence of the perturbation produced by the sphere upon the uniform field at all points outside the sphere?

(III-1) (20 points) A classical electron moves in a circular orbit around a proton. Derive a differential equation for the electron energy, taking into account the classical radiation loss. On this basis calculate the approximate time it takes for a weakly bound (almost free) electron to fall into the first Bohr orbit.

(III-2) (20 points) What charge distribution gives the spherically symmetric potential $V(r) = e^{-\lambda r}/r$?

(III-3) (10 points) A plane EM wave with an electric field $E = 10^6$ (cgs units) is incident normally on a plane dielectric medium, with dielectric constant $\epsilon = 1.44$. Calculate the pressure exerted by the radiation on the dielectric. (Assume the index of refraction $n = \sqrt{\epsilon}$.)

(III-4) (20 points) A capacitor is made of two concentric cylinders of radius r_1 and r_2 $(r_1 < r_2)$ and length $L \gg r_2$. The region between r_1 and $r_3 = \sqrt{r_1 r_2}$ is filled with a circular cylinder of length L and dielectric constant K (the remaining volume is an air gap).

(a) What is the capacitance?

(b) What are the values of E, P, and D at a radius r in the dielectric $(r_1 < r < r_3)$? In the air gap $(r_3 < r < r_2)$? Assume a potential difference V between r_1 and r_2.

(c) How much mechanical work must be done to remove the dielectric cylinder while maintaining this constant potential difference between r_1 and r_2?

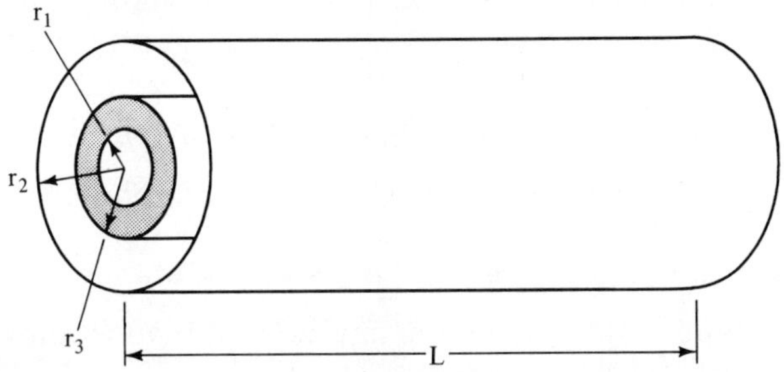

(III-5) (10 points) A coil of N turns is wrapped around an iron ring of radius d and cross section A (d >> A). Assuming a constant permeability $\mu >> 1$ for the iron:

(a) What is the magnetic flux $\Phi = \int B_n dA$ as a function of current I?

(b) If a gap of width δ ($\delta^2 << A$) is cut in the ring, what is the flux for the same current I?

(c) What is the field energy in the iron? In the gap?

(d) With such a gap in the ring, what is the self-inductance?

(III-6) (20 points) A very small circular loop of radius a is initially coplanar and concentric with a much larger circular loop of radius b ($a << b$). A constant current I is passed in the large loop, which is kept fixed in space, and the small loop is rotated with angular velocity ω about a diameter. The resistance of the small loop is R, and its inductive reactance is negligible.

(a) Calculate the current in the small loop as a function of time.

(b) Calculate how much torque must be exerted on the small loop to rotate it.

(c) Calculate the induced emf in the large loop as a function of time.

(IV-1) (5 points) Calculate the electrostatic energy of three charges q, q, and -q located at the vertices of an equilateral triangle of side a.

(IV-2) (5 points) State within a few powers of ten the ratio of electrical to gravitational force between a proton and an electron.

(IV-3) (5 points) A point charge q is at a distance d from a conducting plane. How much energy is required to move the charge infinitely far from the plane?

(IV-4) (5 points) A uniform electric field E_o in the x-direction is produced by an appropriate charge configuration. A thin sheet of charge σ per unit area is placed perpendicular to the x-direction at $x = 0$. If the initial charge configuration is assumed to be undisturbed by the presence of the sheet, what is the total electric field on each side of the sheet?

(IV-5) (5 points) Consider two concentric spherical metal shells of radii r_1 and r_2 ($r_2 > r_1$). If the outer shell has a charge q and the inner shell is grounded, what is its charge?

(IV-6) (5 points) What is the magnetic field inside a long, straight, uniform wire of radius R which is carrying a current I?

(IV-7) (5 points) Two identical iron toroids are wound with N and 2N turns of identical wire, respectively. Assume that the 2N turns requires exactly twice the wire length of the N turns. If the toroids are connected in series, what is the ratio of the potentials across the two windings when:

(1) direct current flows in the windings? __________

(2) high-frequency alternating current flows? __________

(IV-8) (5 points) A long, thin wire carrying a current I is placed at a distance d from a semiinfinite slab of soft iron. The wire is parallel to the surface of the iron. If we assume the iron to have infinite permeability (i.e., $\mu = \infty$), what is the force per unit length on the wire? State whether it is repulsive or attractive.

(IV-9) (5 points) A small, uniformly magnetized bead of volume V is located at the center of a circular loop of radius r carrying a current I. If the magnetic moment per unit volume of the bead is M, directed parallel to the plane of the loop, what is the torque acting on the loop?

(IV-10) (5 points) The two rails of a railroad track are insulated from each other and from ground and are connected by a millivoltmeter. What is the reading when a train travels 180 km/hr down the track, assuming that the vertical component of the earth's field is 0.2 gauss and that the tracks are separated by one meter?

(IV-11) (5 points) An electric dipole, m, is located in a region of constant electric field, E, at an angle α to the field. How much work is required to rotate the dipole 180° about an axis perpendicular to m?

(IV-12) (5 points) A very long, thin rod of dielectric constant K is placed in a homogeneous field E_o parallel to the direction of the field. What are the values of E and D in the interior of the rod?

(IV-13) (5 points) What is the relation between R, L, and C for critical damping in a series LCR circuit?

(IV-14) (5 points) The average light intensity on the earth's surface is 1.3×10^3 joule/m^2/sec. What are the peak values of the E and B fields in volts/m and w/m^2 assuming that the light is monochromatic?

(IV-15) (5 points) What are the boundary conditions for the electric field vector at the interface of two dielectrics, when a surface charge of density δ is present at the interface?

(IV-16) (5 points) How does the potential of an electric dipole depend on the distance r from the dipole?

(IV-17) (5 points) Give a rough estimate of the magnetic field strength at the surface of the earth.

(IV-18) (5 points) What is the electric potential inside an isolated conducting spherical shell of radius R carrying a charge Q?

(IV-19) (5 points) Give expressions for the energy density and momentum density of electromagnetic fields in vacuum.

(IV-20) (5 points) Give the magnitude and direction of the Poynting vector at the surface of a long straight wire of circular cross-section carrying a direct current I. The radius of the wire is b, and the resistance per unit length is R.

(V-1) (5 points) Consider a parallel LC circuit operated at a frequency ω below its resonant frequency ω_o. Is its reactance capacitive or inductive?

(V-2) (5 points) What is the force on an electric dipole of strength P in a uniform electric field E?

(V-3) (5 points) Suppose N identical capacitors are connected in parallel to a potential difference V. What is the potential difference obtained when these capacitors are reconnected in series, their charges being left undisturbed?

(V-4) (5 points) Two identical coils, each of self-inductance L, are connected in series and placed so close to each other that all the flux from one coil links the other. What is the total self-inductance of the system?

(V-5) (5 points) An isolated metallic object is charged in vacuum to a potential V_o, its electrostatic energy being W_o. It is then disconnected from the source of potential, its charge being left unchanged, and is immersed in a large volume of dielectric, with dielectric constant K. What is its electrostatic energy?

(V-6) (5 points) In a spherical electromagnetic wave, how does the magnitude of the electric vector depend upon the distance r from the source, for large values of r?

(V-7) (5 points) An electromagnetic wave is normally incident upon a perfectly conducting surface. In the reflected wave, has either the E or the H vector shifted 180° in phase, and, if so, which one?

(V-8) (5 points) Consider two charged metallic spheres, each of radius R, separated by a distance d ($d > 2R$). One sphere carries a charge +Q and the other a charge -Q. Is the force between the spheres greater than, equal to, or less than the force between two point charges +Q and -Q, separated by a distance d?

(V-9) (5 points) Give an expression for the force per unit length between two long, parallel wires separated by a distance d, if the wires carry equal currents I flowing in the same direction. State whether the force is attractive or repulsive.

(V-10) (5 points) A laboratory C magnet has a pole diameter of 15 cm and an air gap of 1 cm. The iron path is 1 meter. The magnet is wound with 22,000 turns of copper wire.

(a) Taking the initial permeability of iron to be 1000, compute the induction per ampere in the gap in webers/m^2-amp.

(b) For currents above 0.5 amperes the magnet appears to saturate. Using the B-H curve given below graphically estimate the induction for a current of 1 ampere.

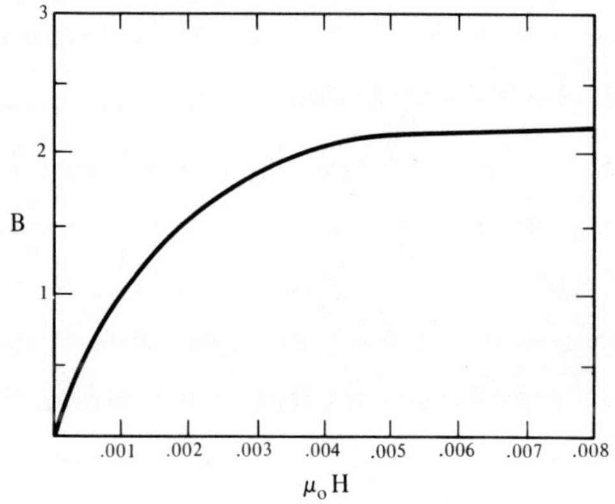

(V-11) (20 points) Consider the simple system shown in the diagram to transform electrical into mechanical energy. Two long parallel guide wires, of zero resistance, separated by a distance ℓ, are connected to a potential difference ϵ. A bar of resistance R makes contact with these wires, and can slide parallel to itself, always remaining perpendicular to the wires. An externally applied uniform magnetic field B is perpendicular to the plane of the bar and the wires.

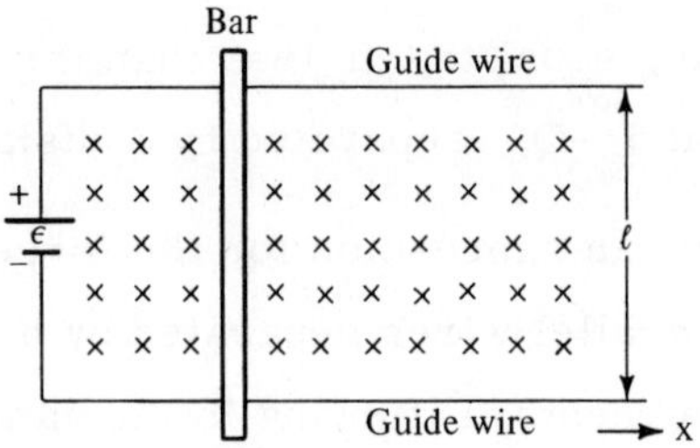

(a) If there is no external mechanical load, what is the steady-state velocity reached by the bar?

(b) If the mass of the bar is m, obtain an expression for the velocity of the bar as a function of time t, assuming that it starts from rest at $t = 0$.

(c) If we apply a constant external force F opposite to the direction of motion of the bar, what is its new steady-state velocity?

(d) Under the conditions of (c), what is the efficiency of our machine, i.e., what fraction of the energy supplied by the battery is converted to mechanical work?

(V-12) (20 points) A parallel-plate capacitor consists of two circular plates of radius r separated by a small gap d ($d \ll r$). Charges $+Q_o$ and $-Q_o$ are placed on the two plates and, at time $t = 0$, their centers are connected with a thin, straight wire of resistance R. Assume that R is very large, so that at any time, the field across the plates remains uniform, and inductance can be neglected.

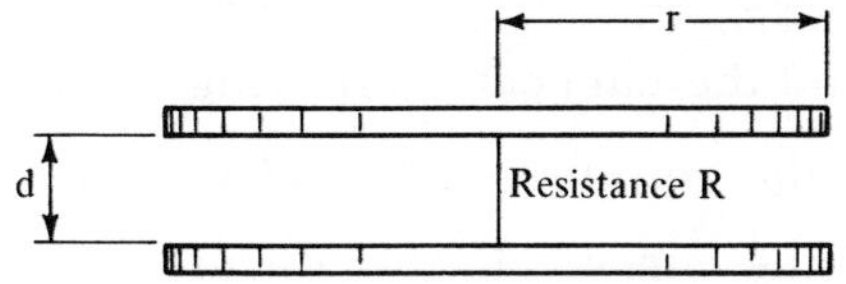

(a) Calculate the charge on each capacitor plate as a function of time.

(b) Calculate the total current crossing a ring of radius ρ ($\rho < r$) in either of the plates, as a function of time. (The ring ρ is concentric with the edge of the plate.)

(c) Calculate the magnetic field between the plates as a function of time and radial distance from the center.

(d) Explain in detail why the only nonvanishing component of the magnetic field is in the azimuthal direction.

(V-13) (10 points) A light wave has a frequency of 4×10^{14} cycles/second and a wavelength of 5×10^{-7} meters. What is its speed? What is the index of refraction of the medium in which it is travelling? What is its wavelength after it passes from the medium into air?

(VI-1) (20 points) The system shown in the diagram consists of two flat conducting strips of length ℓ, width b (perpendicular to the plane of the diagram), separated by a small gap a ($a \ll b, \ell$). The right ends of the strips are shorted, and a battery of voltage V_o is connected across the left ends. The current is assumed to flow only parallel to the ℓ-dimension of the strips. Neglect all resistances and all effects arising from the finite speed of propagation of electromagnetic fields.

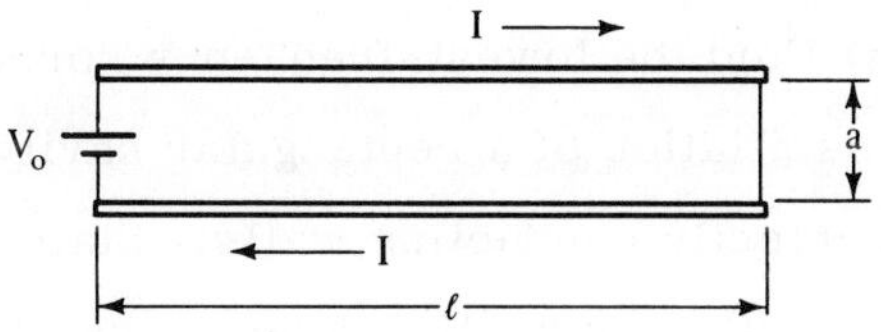

(a) What is the relation between the magnetic field B between the strips and the current I flowing in the circuit?

(b) What is the self-inductance of the circuit?

(c) What is the current in the circuit as a function of time?

(d) What is the voltage across the strips as a function of the distance x from the shorted end?

(e) What is the rate of flow of energy down the system as a function of distance from the shorted end?

(VI-2) (20 points) Find the torque and the force between two circular loops of wire, carrying the same currents I, and of the same radius R, when they are located a distance L apart, with $L \gg R$, and with their axes parallel and the currents in the same direction. Express the torque and the force in terms of the angle θ between their axes and their line of centers.

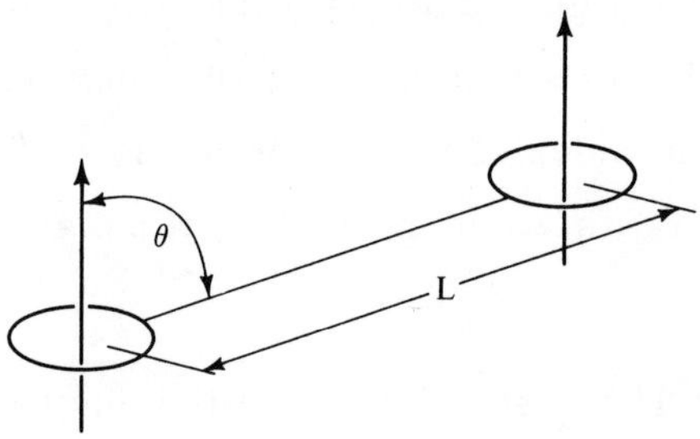

(VI-3) (10 points) A long wire is bent into the hairpin-like shape shown in the figure. Find an exact expression for the magnetic field at the point P which lies at the center of the half-circle:

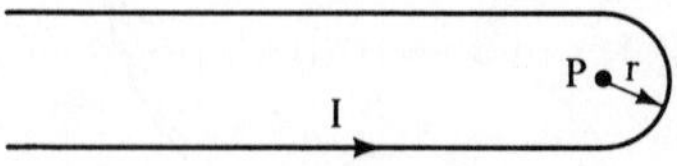

(VI-4) (20 points) Find the lowest-frequency normal-mode electromagnetic oscillation of a rectangular cavity resonator of sides $a > b > d$, with perfectly conducting walls. State the resonant frequency, and describe the spatial dependence of the fields.

(VI-5) (20 points) Consider a plane electromagnetic wave of frequency ω normally incident on a nonmagnetic metallic surface with given conductivity σ.

(a) Write down the partial differential equation for the magnetic field, appropriate to the interior of the metal. Assume that ω is small enough that displacement current effects can be neglected.

(b) State the boundary conditions for the tangential components E^t and H^t of the electric and magnetic fields at the surface.

(c) Evaluate the (complex) surface impedance $Z(\sigma, \omega)$ defined by $\vec{E}^t = Z\vec{H}^t \times \hat{n}$ where $\hat{n}$ is a unit vector normal to the surface.

(VI-6) (10 points) A straight circular cylindrical metal wire of uniform conductivity σ and cross-sectional area A carries a steady current I. Determine the direction and magnitude of the Poynting vector at the surface of the wire. Integrate the normal component of the Poynting vector over the surface of the wire for a segment of length L and compare your result with the Joule heat produced in this segment.

(VII-1) (20 points)

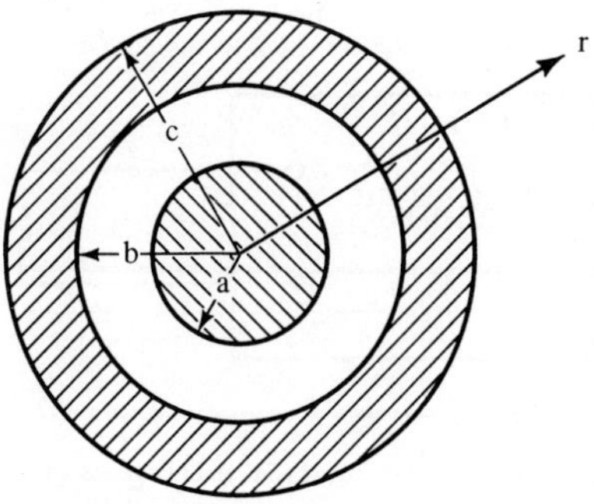

(a) A long coaxial cable (shown in cross section) has a uniform current I flowing in the center conductor into the paper, and the same current I flowing in the outer cylinder out of the paper. Find the magnetic field $\vec{\beta}$ in each of the four regions.

(i) $r < a$

(ii) $a < r < b$

(iii) $b < r < c$

(iv) $r > c$

(b) Calculate the self-inductance L of a 10-cm length of this same coaxial cable. (Assume $c >> a$ and $c >> c - b$ so that we can neglect the thickness of the conductor.)

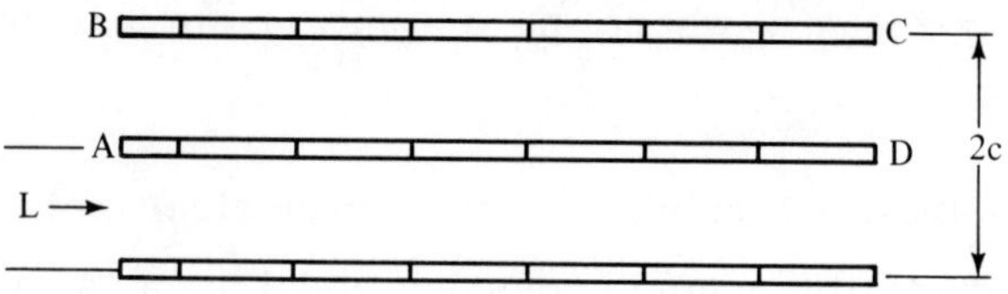

(VII-2) (20 points)

(a) Given a black box containing unknown emf's and resistances connected in an unknown way such that (1) a 10-ohm resistance connected across its terminals draws a current of one ampere, and (2) an 18-ohm resistance draws only 0.6 amp. What resistance will draw 0.1 amp?

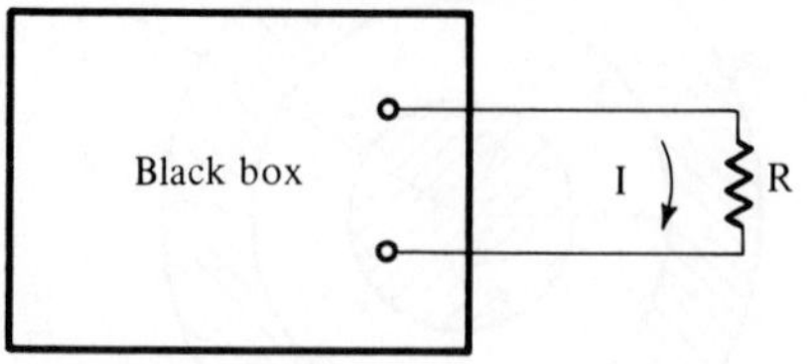

(b) Draw a simple filter which will greatly attenuate 60-cps ripple voltage from the output of the circuit shown below. Specify exactly the values of circuit elements required.

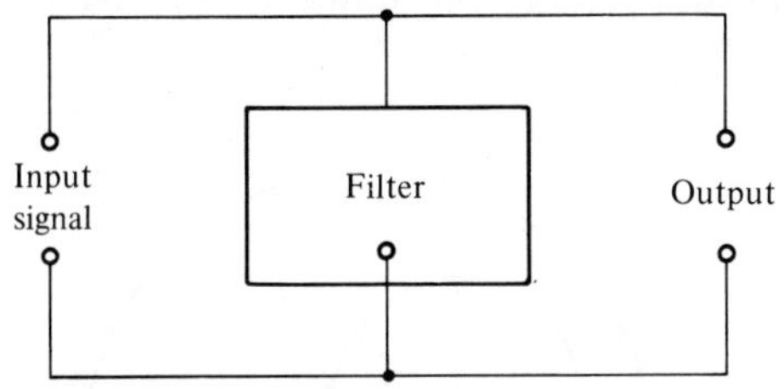

(VII-3) (20 points) Two magnetic dipoles $\vec{\mu}_1$ and $\vec{\mu}_2$ at fixed centers separated by $\vec{r}$ are free to rotate about their centers.

(a) Sketch the configuration for maximum energy and calculate this energy.

(b) Sketch the configuration for minimum energy and calculate this energy.

(VII-4) (10 points)

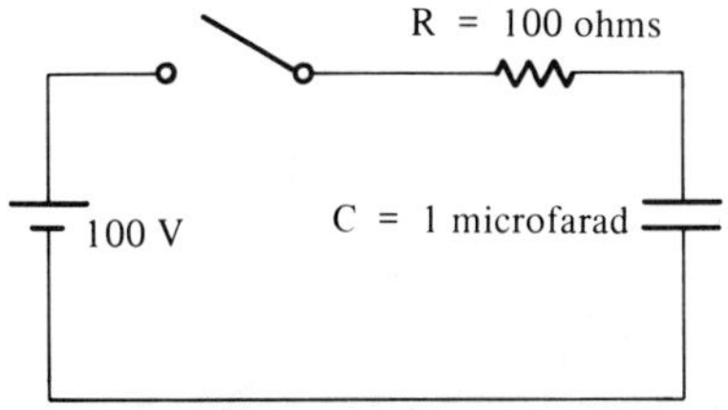

A capacitor C is suddenly connected to a battery of 100 volts through a resistance R. After what time will the capacitor be charged to 50 volts?

t = __________ seconds

(VII-5) (10 points) An electron is launched with a velocity $v = 10^4$ cm/sec at 45° to a uniform field $H = 10^4$ oersteds. Describe quantitatively and completely the ensuing motion in one sentence. (Be sure to specify completely and quantitatively the exact motion.)

(VII-6) (20 points) A long straight wire of radius a has a circular hole of radius b parallel to the axis of the wire but displaced from the center by a distance c. A current I flows in the wire and is uniformly distributed across the conductor. Find the magnetic field everywhere in space.

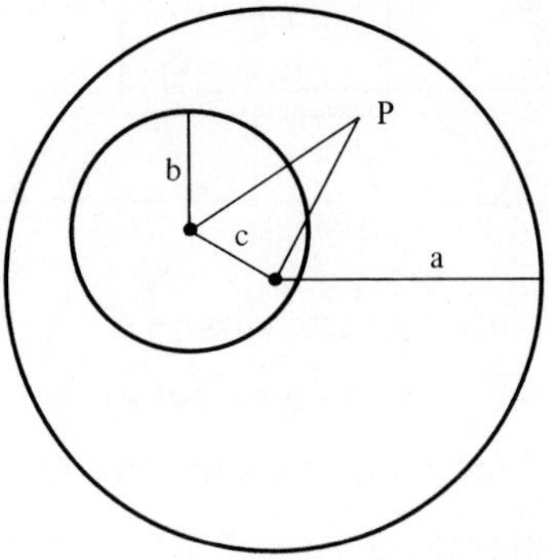

$b + c < a$

(VIII-1) (20 points)

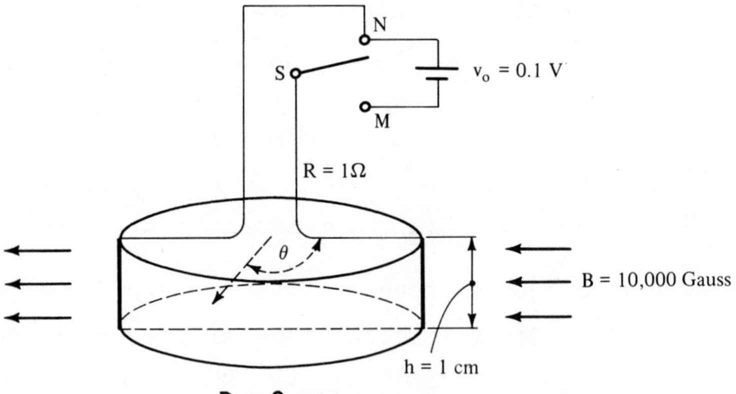

D = 2 cm

A light brass wire of total resistance $R = 1$ ohm is wound around a disc of plastic (density $\rho = 1\ gm/cm^3$) and suspended so as to make an essentially undampened torsion pendulum of period $T = 10$ sec. The disc is initially at rest in a uniform magnetic field, $B = 10,000$ gauss, with $\vec{B}$ in the plane of the loop, i.e., $\theta = 90°$. At time $t = 0$, switch S is switched to M and then at $t = T_1 = 10^{-4}$ sec switched to N. Find the amplitude (in radians) of the ensuing oscillations as a function of t.

(VIII-2) (20 points) A box provided with two terminals is known to contain an inductance of negligible resistance, a capacitor, and a resistor. When 100 volts dc is connected to the terminals, a current of 0.1 amp flows. When 100 rms volts ac at 60 cycles/sec is connected, 1 amp rms flows. If the frequency is increased and the applied voltage maintained constant, the current rises to a very high maximum at 1000 cycles/sec. How are the three components connected inside the box, and what values do they have?

(VIII-3) (20 points) Starting with Maxwell's equations, obtain an expression describing the propagation of a plane wave of frequency ω in an extended medium of conductivity σ, dielectric constant ϵ, and magnetic permeability μ.

(VIII-4) (10 points) What minimum energy must a proton have in order to produce a proton-antiproton pair upon striking a deuteron?

__________ MeV.

(VIII-5) (10 points) An inductance is suddenly connected to a 6-volt battery through a resistance R. What is the steady-state current drawn from the battery? After what time will the battery be delivering one-half its steady-state current?

(VIII-6) (20 points) Write expressions for the magnetic scalar and vector potentials from which the following magnetic field can be derived:

$$\vec{B} = k(y\hat{i} + x\hat{j}).$$

(IX-1) (15 points) The electrons in a long tube of completely ionized hydrogen gas are flowing along the tube with mean velocity 10^5 cm/sec in a circular beam of diameter 50 cm. The total beam current is 10^4 amps. Find the magnitude and sense of the force F on one electron at the edge of the beam.

(IX-2) (20 points) The effective conductivity of a region of space with N electrons per cubic meter is $\sigma = -i(Ne^2/\omega m)$, where e is the electron charge and m the electron mass. From Maxwell's equations, derive the velocity of propagation of electromagnetic waves in this space and from this the index of refraction. Explain how this problem is connected with the reflection of radio waves from the upper atmosphere.

(IX-3) (10 points) The network

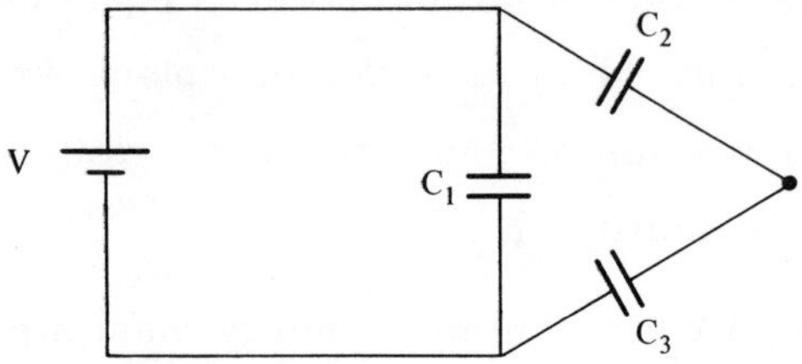

is raised to a potential difference of V volts. Find the electrical energy stored in the network.

(IX-4) (5 points) To within two orders of magnitude, what is the pressure in mm of Hg in a vacuum tube?

(IX-5) (20 points) A rectangular hoop of conductor has height H and width W. At $t = 0$ it is dropped from rest. At that time the bottom edge of the hoop is a height h above the plane $y = 0$. Above that plane there is no magnetic (or electric) field. Below that plane there is a uniform magnetic field B perpendicular to the plane of the drawing and directed out of paper. The hoop has mass m, resistance R. Find the motion of the hoop for all times. Find the velocity v for all times, and plot it versus t. Of special importance are times $t = 0$ to t_1, t_1 to t_2, and t greater than t_2.

How is the motion affected if the hoop is made bigger but retains the same ratio W/H, and it is made of the same material as before.

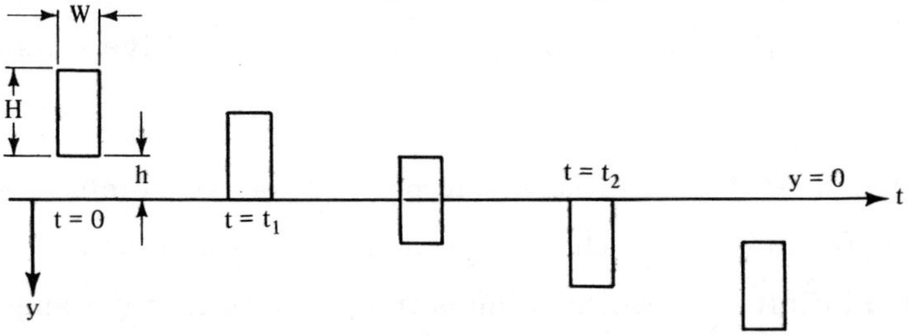

(IX-6) (10 points) A point charge q (coulombs) is situated a distance d from an infinite conducting grounded plate. Find the charge density at the surface of the plate as a function of the distance from the normal to the plate from the charge.

(IX-7) (20 points) The plates of a parallel-plate condenser are separated by 1 cm. The potential difference across the plates is 1000 volts. The plates are square and 1 m on a side.

(a) Calculate the force between the plates in newtons.

(b) Calculate the charge per unit area in coulomb/m^2.

(Neglect edge effects.)

(X-1) (20 points) Suppose that one measures the electrical conductivity of the following materials at room temperature:

high purity copper

n-type germanium

niobium

One then plunges each of these samples into liquid helium (4° K) and repeats the measurement. How much does the conductivity of each material change? (qualitative answer) In which direction? Why?

(X-2) (20 points) A very thin hollow cylinder carrying a current is 10 cm in diameter. It is surrounded by a hollow cylinder 20 cm in diameter and concentric with it.

(a) Calculate the self-inductance per unit length if this outer cylinder carries an equal current in the opposite direction and the two are part of an electric circuit.

(b) State whether the force on the outer cylinder is one tending to burst apart or to collapse the cylinder. (Give reasons for your answer.)

(X-3) (20 points) Find the electric potential at all points in space produced by an electric charge Q located at a distance d from two grounded infinite conducting planes intersecting at right angles.

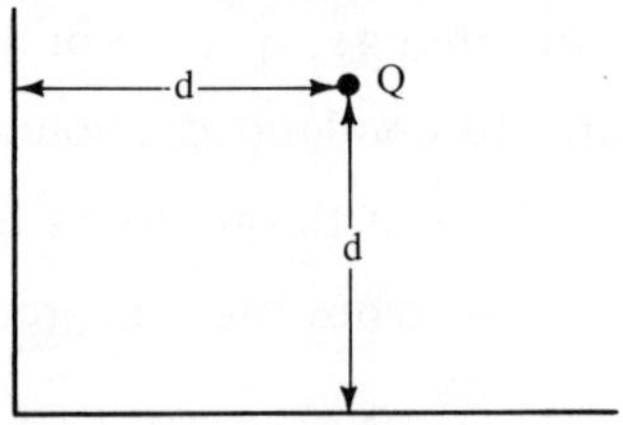

(X-4) (20 points) A toroidal ring is made from a bar 1 cm in diameter and 1 m long bent into a circle. It is wound with 100 turns per cm. If the permeability of the bar is that of free space, calculate:

(a) The magnetic field inside the bar when 100 amps are circulating through the turns.

(b) Calculate the self-inductance of the coil in henries assuming that the coil is made of very thin strips and is wound tightly on the bar.

(c) Calculate the electrical energy necessary to build up the field as the current is raised to full value.

(d) Calculate the energy stored in the magnetic field from the values of B and H and the volume.

Neglect the change in field with radius, and assume it to be uniform and equal to that at the center of the bar. State all units.

(X-5) (20 points) Two unequal condensers (C_1, C_2) are charged separately to the same potential difference (V), and subsequently the positive terminal of one is connected to the negative terminal of the other. Then the two outermost connections are shorted together.

(a) Calculate the final charge on each condenser.

(b) Calculate the loss in electrostatic energy.

SOLUTION SET OF
ELECTRICITY AND MAGNETISM

(I-1)

a. The skin depth is defined by

$$\delta \sim \frac{c}{\sqrt{(2\pi\mu\omega\sigma)}}$$

which is proportional to the inverse of the square root of the frequency. Therefore we have

$$\frac{\delta_1}{\delta_2} = \sqrt{\frac{\omega_2}{\omega_1}} = \sqrt{\frac{(10^8)}{1000}} \sim 300.$$

b. The energy density of the laser is related to the peak value of the E-field by:

$$\text{E. D.} = \frac{1}{8\pi} E^2$$

from which we find

$$E = \sqrt{(8\pi \times 10^{13})} = 1.6 \times 10^7 \text{ cgs units}$$

$$= 4.8 \times 10^{11} \frac{\text{volts}}{\text{m}}$$

c. The circuit equation is

$$L\frac{dI}{dt} + RI + \frac{Q}{C} = 0. \tag{1}$$

Differentiating (1) with respect to t, we get

$$L\frac{d^2I}{dt^2} + R\frac{dI}{dt} + \frac{I}{C} = 0. \tag{1)'$$

Substituting

$$I = A_o e^{-i\delta t} \tag{2}$$

into (1)' we obtain

$$L\delta^2 + iR\delta - \frac{1}{C} = 0.$$

The solutions of δ are

$$\delta = \frac{-iR \pm \sqrt{(-R^2 + 4L/C)}}{2L} .$$

The condition of critical damping is

$$R^2 - \frac{4L}{C} = 0.$$

(I-2) Let $x = A \sin \omega t$. (1)

The energy of the oscillating electron with mass m is

$$E = \frac{1}{2} m A^2 \omega^2 ,$$

from which we get

$$A = \sqrt{\frac{2E}{m\omega^2}} . \tag{2}$$

The rate at which the energy is radiated is

$$\frac{dE}{dt} = - \frac{2}{3c^3} e^2 a^2 \tag{3}$$

where $a = -A\omega^2 \sin \omega t$. Substituting (2) into (3) and averaging over one period, we get

$$\frac{dE}{dt} = - \frac{2}{3} \frac{e^2}{c^3} \frac{1}{2} \omega^4 \frac{2E}{m\omega^2} = - \frac{2}{3} \frac{e^2 E \omega^2}{mc^3} . \tag{4}$$

The energy of the electron as a function of t is

$$E = E_o \exp(-\frac{2e^2\omega^2}{3mc^3})t. \qquad (5)$$

a. $$\Delta E \approx \Delta t < \frac{2}{3c^3}e^2a^2 > = \frac{2\pi}{\omega}\frac{A^2}{3c^3}e^2\omega^4$$

$$= \frac{2\pi A^2 e^2 \omega^3}{3c^3} = \frac{10^{-16} \times 2\pi(4.8)^2 \times 10^{-20} \times 10^{45}}{3 \times 27 \times 10^{30}}$$

$$= 1.8 \times 10^{-21} \text{ ergs} = 1.1 \times 10^{-9} \text{ eV}.$$

b. The mechanical average energy of the electron is the sum of the mean kinetic energy and the mean potential energy. We find

$$E = \frac{1}{2}mv^2 = \frac{1}{2}mA^2\omega^2,$$

where $v = A\omega$ is the maximum velocity of the electron. The ratio of the radiated energy per cycle to the mechanical average energy is

$$\frac{\Delta E}{E} = \frac{2\pi}{\omega}\frac{A^2}{3c^3}e^2\omega^4\frac{2}{mA^2\omega^2} = \frac{4\pi e^2\omega}{3c^3m}$$

$$= \frac{4\pi \times (4.8)^2 \times 10^{-20} \times 10^{15}}{3 \times 27 \times 10^{30} \times 9.1 \times 10^{-28}}$$

$$= 3.9 \times 10^{-8}.$$

c. Taking $E = \frac{1}{2}E_o$, we can solve for time τ from (5).

$$\tau = (\ln 2)\frac{3mc^3}{2e^2\omega^2}$$

$$= \frac{0.69 \times 3 \times 27 \times 10^{30} \times 9.1 \times 10^{-28}}{2 \times (4.8)^2 \times 10^{-20} \times 10^{30}}$$

$$\approx 1.1 \times 10^{-7} \text{ sec}.$$

(I-3)

a. The potential of a single charge as a function of distance r is

$$V = \frac{q}{r}. \tag{1}$$

Differentiating (1) with respect to $-x$, we get the potential of a dipole orienting in the x-axis direction,

$$V_d = -\frac{dV}{dx}\Delta x = \frac{xq\Delta x}{r^3} = \frac{P\cos\theta}{r^2}$$

where $P = q\Delta x$ is the dipole moment. The E-field of the dipole is

$$\vec{E}_d = \frac{P\sin\theta}{r^3}\hat{\theta}_1 + \frac{2P\cos\theta}{r^3}\hat{r}_1$$

where $\hat{\theta}_1$ and $\hat{r}_1$ are unit vectors. The potential energy of two parallel dipoles is

$$W = -\vec{P}\cdot\vec{E}_d = \frac{P^2}{d^3}$$

where we have used $\theta = 90°$ and $r = d$.

b. The dipole moment of a conducting sphere in a uniform field E is known to be R^3E, where R is the radius of the sphere. The force between the two spheres is $-\frac{dW}{dr}$, or

$$F = (ER^3)^2\frac{3}{d^4} = \frac{3E^2R^6}{d^4} \quad \text{(repulsive)}.$$

(I-4)

a. According to Ampere's law we have

$$\oint \vec{B}\cdot d\vec{S} = \frac{4\pi}{c}I$$

where I is the net current passing through the area defined by the closed curve. Using $I = \rho v\pi r^2$ and the fact that B is independent of angles we obtain

$$B(2\pi r) = \frac{4\pi}{c} \rho v \pi r^2$$

or

$$B = \frac{2\pi\rho v r}{c}.$$

b. Let

$$\delta = \frac{2\pi\rho v}{c}.$$

The force acting on a proton at $r = y_o$, moving with velocity V' is

$$F = m\frac{dv_y}{dt} = -\frac{eV'B}{c} = \frac{-e\delta y_o V'}{c}$$

or

$$\frac{dv_y}{dt} = -\frac{e\delta}{cm} y_o V'.$$

After integrating we obtain

$$v_y = -\frac{e\delta}{cm} y_o V't \tag{1}$$

and

$$\Delta y = -\frac{e\delta}{2cm} y_o V't^2 = -\frac{e\delta}{2cm} y_o V'(\frac{L}{V'})^2$$

$$= -\frac{e\delta y_o L^2}{2cV'm}$$

which is negligible compared with y_o when $L << \text{focal length}$. The direction of the proton after passing through the cylinder is defined by

$$\frac{v_y}{V'} = \frac{y_o}{\text{focal length}}. \tag{2}$$

Substituting (1) into (2) we obtain

$$\text{focal length} = y_o\left(\frac{cmV'}{e\delta y_o V't}\right) = \frac{cm}{e\delta t} = \frac{c^2 m}{e2\pi\rho v\frac{L}{V'}} = \frac{c^2 mV'}{2\pi\rho vLe}$$

$$= \left(\frac{c^2 m}{2\pi e\rho L}\right)\frac{V'}{v},$$

where we have used $t = L/V'$ and $\delta = 2\pi\rho v/c$.

(I-5)

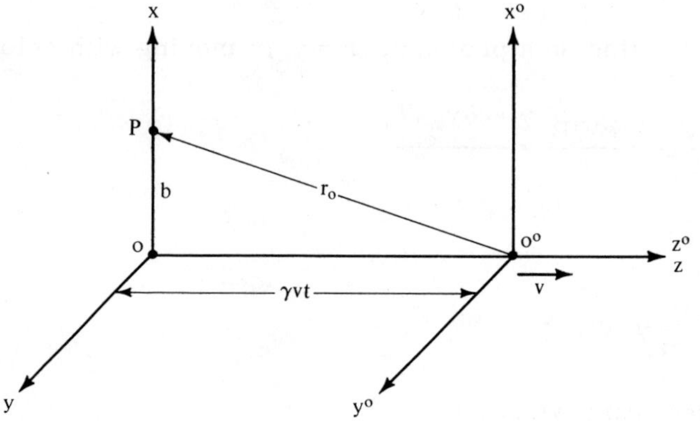

Let oxyz be the laboratory system and $o^\circ x^\circ y^\circ z^\circ$ be the system moving with the charge. The potential of a point charge at the point P in the rest frame of the charge is

$$A_x^o = A_y^o = A_z^o = 0$$

and

$$V^o = \frac{e}{r_o} \text{ with } r_o = (x_o^2 + z_o^2)^{1/2} = (b^2 + \gamma^2 v^2 t^2)^{1/2}$$

$$\text{and } \frac{\partial r_o}{\partial z_o} = -\frac{z_o}{r_o}, \quad \frac{\partial r_o}{\partial t} = \frac{\gamma^2 v^2 t}{r_o} \tag{1}$$

The Lorentz transformation from the rest frame of the charge to the laboratory system is:

$$T = \begin{pmatrix} \gamma & -i\gamma\beta \\ i\gamma\beta & \gamma \end{pmatrix}$$

where we have used the abbreviations

$$\beta = \frac{v}{c} \quad \text{and} \quad \gamma = \frac{1}{\sqrt{1 - v^2/c^2}}.$$

With the above transformation matrix, we obtain the following relation between the potential in the moving system and the potential in the rest system:

$$\begin{pmatrix} A_x \\ A_y \\ A_z \\ iV \end{pmatrix} = \begin{pmatrix} 1 & 0 & 0 & 0 \\ 0 & 1 & 0 & 0 \\ 0 & 0 & \gamma & -i\gamma\beta \\ 0 & 0 & i\gamma\beta & \gamma \end{pmatrix} \begin{pmatrix} 0 \\ 0 \\ 0 \\ iV^o \end{pmatrix}$$

or

$$A_x = 0; \quad A_y = 0$$

$$A_z = \gamma\beta V^o = \frac{\gamma\beta e}{r_o}$$

and

$$iV = i\gamma V^o = \frac{i\gamma e}{r_o}. \tag{2}$$

The Lorentz transformation which connects coordinates in the laboratory system to those in the rest system is

$$z_o = \gamma(z - vt), \quad x_o = x, \quad y_o = y$$

and

$$t_o = \gamma(t - \frac{v}{c^2} z)$$

from which we find the following relations:

$$\frac{\partial}{\partial z} = \frac{\partial}{\partial z_o}\frac{\partial z_o}{\partial z} + \frac{\partial}{\partial t_o}\frac{\partial t_o}{\partial z}$$

$$= \gamma\frac{\partial}{\partial z_o} - \gamma\frac{v}{c^2}\frac{\partial}{\partial t_o}$$

$$\frac{\partial}{\partial t} = \frac{\partial}{\partial z_o}\frac{\partial z_o}{\partial t} + \frac{\partial}{\partial t_o}\frac{\partial t_o}{\partial t}$$

$$= -\gamma v\frac{\partial}{\partial z_o} + \gamma\frac{\partial}{\partial t_o}$$

or

$$\frac{\partial}{\partial z} + \frac{\beta}{c}\frac{\partial}{\partial t} = \gamma(1 - \beta^2)\frac{\partial}{\partial z_o} = \frac{1}{\gamma}\frac{\partial}{\partial z_o}. \tag{3}$$

Using the relations

$$\vec{E} = -\nabla V - \frac{1}{c}\frac{\partial \vec{A}}{\partial t}$$

and

$$\vec{B} = \nabla \times \vec{A}$$

we find the fields at P of a point charge in uniform motion are

$$E_z = -\frac{\partial V}{\partial z} - \frac{1}{c}\frac{\partial A_z}{\partial t} = \frac{\gamma e}{r_o^2}\left(\frac{\partial}{\partial z} + \frac{\beta}{c}\frac{\partial}{\partial t}\right)r_o = \frac{e\gamma^2}{r_o^2}(1 - \beta^2)\frac{\partial r_o}{\partial z_o}$$

$$= \frac{-\gamma evt}{r_o^3} = -\frac{\gamma\, evt}{(b^2 + \gamma^2 v^2 t^2)^{3/2}}$$

$$E_x = -\frac{\partial V}{\partial x} = \frac{b\gamma e}{r_o^3} = \frac{\gamma eb}{(b^2 + \gamma^2 v^2 t^2)^{3/2}}; \qquad E_y = 0$$

and

$$B_y = -\frac{\partial A_z}{\partial x} = \frac{\gamma\beta eb}{r_o^3} = \frac{\gamma ebv}{(b^2 + \gamma^2 v^2 t^2)^{3/2}}; \qquad B_x = B_z = 0$$

where we have used the relations

$$r_o^2 = b^2 + \gamma^2 v^2 t^2,$$

$$\frac{\partial r_o}{\partial z_o} = \frac{z_o}{r_o} = \frac{-\gamma vt}{r_o} \quad \text{and Eq. (3).}$$

(II-1) Let

$$f(r) = \nabla^2 \frac{1}{|\vec{r}|}$$

and integrate over dV,

$$I = \int f(r)dV = \int \nabla^2 \frac{1}{r} dV$$

$$= \iint_S \nabla \frac{1}{r} \cdot d\vec{s} \quad \text{(Gauss's theorem)}$$

where S is the surface which encloses the volume V. In spherical coordinates the integral becomes:

$$I = -\iint_S \frac{1}{r^2} r^2 d\theta\, d\varphi \sin\theta = -\iint_S d\theta\, d\varphi \sin\theta$$

$= 0$ if the point $r = 0$ is not inside surface S.

$= -4\pi$ if the point $r = 0$ is inside surface S.

We see that the function $f(r)$ satisfies the definition of the generalized function $-4\pi\delta(\vec{r})$. Therefore

$$f(r) = \nabla^2 \frac{1}{|\vec{r}|} = -4\pi\delta(\vec{r}).$$

(II-2) The electrostatic field, E_o, due to a charge q is

$$E_o = \frac{q}{r^2}.$$

At large distance the field, E_1, of a dipole is

$$E_1 = -\frac{dE_o}{dx}\Delta x$$

where Δx is the separation of the two charges. To find the field, E_2, due to two dipoles with opposite sense (quadrupole) separated by a distance Δy, we need simply differentiate the dipole field, E_1, with respect to $-y$. Therefore,

$$E_2 = \frac{d^2E_o}{dy\,dx}\Delta x\Delta y.$$

Similarly, the field, E_3, due to the eight charges at the corners of a cube (octopole) is just

$$E_3 = -\frac{d^3E_o}{dz\,dy\,dx}\Delta x\Delta y\Delta z,$$

from which we find E_3 is proportional to r^{-5} since $\Delta x = \Delta y = \Delta z = \text{constant}$.

(II-3) In the radiation zone, the electric field and the magnetic induction for dipole radiation are:

$$\vec{E} = k^2(\vec{n} \times \vec{p}) \times \vec{n}\,\frac{e^{ikr}}{r}$$

and

$$\vec{B} = k^2(\vec{n} \times \vec{p})\frac{e^{ikr}}{r}.$$

Here $k = 2\pi/\lambda$ is the wave number, $\vec{p}$ is the dipole moment, and $\vec{n}$ is the unit vector in the direction of propagation which equals the unit position vector here. (Since the intensity obeys the inverse square law, the field has to be proportional to $1/r$. The direction is defined by the relation $\vec{E} \times \vec{B} // \vec{n}$. The expansion of the vector potential gives one k factor for the term corresponding to dipole radiation. The second k factor simply comes from the relation $\vec{B} = \nabla \times \vec{A} = i\vec{k} \times \vec{A}$). The fields due to the two dipoles are

$$\vec{E}' = -\frac{\lambda}{2}\frac{d\vec{E}}{dx}$$

$$\simeq -\frac{i\lambda k^3}{2}(\vec{n} \times \vec{p}) \times \vec{n}\,\frac{xe^{ikr}}{r^2}$$

and

$$\vec{B}' = -\frac{\lambda\, d\vec{B}}{2dx}$$

$$\simeq -\frac{i\lambda k^3}{2}(\vec{n} \times \vec{p})\frac{xe^{ikr}}{r^2}$$

where we have neglected all terms proportional to $\frac{1}{r^3}$. The Poynting vector, $\vec{S}$, at large distances is

$$\vec{S} = \frac{c}{4\pi}\vec{E}' \times \vec{B}'^*$$

$$\approx \frac{c\pi k^4 p^2 \cos^2\theta \sin^2\theta \cos^2\varphi}{4r^2}\vec{n}$$

where

$$\cos\theta = \frac{z}{r} \text{ and } \tan\varphi = \frac{y}{x}.$$

(II-4) The resistance of a superconducting loop is very small. If there were any change of the magnetic flux, φ, through the coil, the induced emf would immediately produce a large induced current to keep the magnetic flux constant. Before the two loops are superposed, the magnetic flux through each loop is

$$\varphi_i = AB - CI \tag{1}$$

where C is a constant. After they are put together, the magnetic flux through either loop is

$$\varphi_f = CI' = C(I'_1 + I'_2) \tag{2}$$

where I' is the final total current in the two superposed loops. The condition $\varphi_i = \varphi_f$, therefore, leads to

$$I' = I \quad \text{or} \quad I'_1 = I'_2 = \frac{1}{2}I' = \frac{1}{2}I \tag{3}$$

since the two loops are identical. The initial energy of the system is

$$W_i = \frac{1}{2}LI^2 + \frac{1}{2}LI^2$$

$$= LI^2. \tag{4}$$

The final energy of the system is

$$W_f = \frac{L}{2}I_1'^2 + \frac{L}{2}I_2'^2 + MI'_1I'_2. \tag{5}$$

Furthermore $M = L$ when the two loops are put together. Using (3) and (5), we obtain

$$W_f = \frac{L}{2}I^2.$$

The change of energy is

$$W_f - W_i = -\frac{L}{2}I^2$$

which has been converted into mechanical energy (e.g., work done to the hand or to some friction force) and ratiation energy.

(II-5) The equivalent impedance of the circuit is

$$Z = iL\omega + \frac{1}{\frac{1}{R} + i\omega C}$$

$$= iL\omega + \frac{R - iR^2C\omega}{1 + R^2C^2\omega^2}$$

$$= \frac{R}{1 + R^2C^2\omega^2} + i\left(\frac{L\omega + R^2C^2L\omega^3 - R^2C\omega}{1 + (RC\omega)^2}\right).$$

Using

$$\omega = \frac{1}{\sqrt{(LC)}},$$

we obtain

$$Z = \frac{RL}{L + R^2C} + i\left(\frac{L^2}{L + R^2C}\right)\frac{1}{\sqrt{(LC)}}.$$

The total current through the circuit is:

$$I_{total} = \frac{V}{Z} = \frac{V_o}{Z}e^{i\omega t}.$$

The current through the resistor is:

$$i_R = \frac{1}{1 + i\omega CR}I_{total}$$

$$= V_o e^{i\omega t}\frac{1}{\frac{RL}{L + R^2C} + i\left(\frac{L^2}{L + R^2C}\right)\frac{1}{\sqrt{(LC)}}} \times \frac{1}{1 + i\omega CR}$$

$$= \frac{V_o e^{i\omega t}}{R_1 R_2} e^{-i(\delta_1 + \delta_2)}$$

where

$$R_1 = \left(\frac{RL}{L + R^2C}\right)^2 + \left(\frac{L^2}{L + R^2C}\right)^2 \frac{1}{LC}$$

$$= \frac{L}{L + R^2C}\sqrt{R^2 + \frac{L}{C}}$$

$$R_2 = \sqrt{1 + (\omega CR)^2}$$

$$\delta_1 = \tan^{-1} \frac{1}{R}\sqrt{\frac{L}{C}} \text{ and } \delta_2 = \tan^{-1} \omega CR.$$

(II-6) Let the electric intensity be $\vec{E} = \vec{E}_o e^{i(\omega t - \vec{k}\cdot\vec{r})}$. The displacement current is:

$$J_d = \frac{\epsilon_o}{4\pi}\frac{\partial E}{\partial t} = \frac{i\omega\epsilon_o E}{4\pi}$$

while the conduction current is σE. If the two currents are equal in magnitude, we have

$$\frac{4\pi\sigma}{\omega\epsilon_o} = 1. \tag{1}$$

Therefore we can define a complex index of refraction n

$$n = \frac{1}{c}\sqrt{(\epsilon_o - \frac{4\pi\sigma}{\omega} i)\mu_o} = \sqrt{1 - \frac{4\pi\sigma}{\epsilon_o \omega} i}$$

or using (1) we get

$$n = \sqrt{(1 - i)} \equiv n_o(1 - ik) \tag{2}$$

where k is called the absorption coefficient. The reflection coefficient is related to the indices of refraction at normal incidence by

$$r = \left|\frac{n - 1}{n + 1}\right|^2$$

$$= \frac{(n_o - 1)^2 + (n_o k)^2}{(n_o + 1)^2 + (n_o k)^2}. \tag{3}$$

Both n_o and k are defined in (2) to be real and positive. We find

$$n_o = \sqrt{\frac{1+\sqrt{2}}{2}} \sim 1.10$$

and

$$k = \frac{1}{1+\sqrt{2}} \sim 0.414. \qquad (4)$$

Substituting (4) into (3) we obtain the reflection coefficient $r = 0.047$.

(II-7) An uncharged conducting sphere placed in a uniform electric field is equivalent to an electric dipole situated at the center of the sphere as far as only points outside the sphere are concerned. The electric field of a dipole can be expressed in terms of the two components $\vec{E}_r$ and $\vec{E}_\theta$ as follows

$$\vec{E}_r = \frac{2P\cos\theta}{r^3}\hat{r}_1$$

$$\vec{E}_\theta = \frac{P\sin\theta}{r^3}\hat{\theta}_1$$

where P is the equivalent dipole moment of the sphere;

$$P = R^3 E_o$$

and where R is the radius of the sphere.

(III-1) The rate at which the kinetic energy of an accelerated electron is lost due to radiation is given by

$$-\frac{dE}{dt} = \frac{2a^2e^2}{3c^3} \qquad (1)$$

where E is the total energy of the electron proton system; a is the acceleration of the electron; e is the charge; and c is the velocity of light. The acceleration a is related to the radius, r, of the orbit of the electron through the equilibrium condition

$$ma = \frac{e^2}{r^2} \quad \text{or} \quad a = \frac{e^2}{mr^2}. \tag{2}$$

Using equation (2) and the relation $a = v^2/r$, we find that the total energy of the electron proton system is

$$E = \frac{1}{2}mv^2 - \frac{e^2}{r} = -\frac{e^2}{2r}. \tag{3}$$

Substituting (3) and (2) into (1) we obtain

$$-\frac{e^2}{2r^2}\frac{dr}{dt} = \frac{2e^6}{3m^2r^4c^3}. \tag{4}$$

After obvious cancellation we get

$$\frac{dr}{dt} = \frac{-4e^4}{3m^2r^2c^3}. \tag{5}$$

After integration, (5) becomes

$$T = \int_0^T dt = -\frac{m^2c^3}{4e^4}\int_R^{a_o} 3r^2\,dr = \frac{m^2c^3}{4e^4}(R^3 - a_o^3)$$

where R is the initial distance between the electron and proton and a_o is the Bohr radius.

(III-2) For $r \neq 0$ the Poisson's equation in spherical coordinates is

$$\nabla^2 V \equiv \frac{1}{r}\frac{d^2}{dr^2} rV(r) + \Lambda V(r) = -4\pi\rho(r) \tag{1}$$

where Λ is a differential operator depending on the angles. For a spherically symmetric potential, $\Lambda V(r) = 0$. Using the given expression for $V(r)$ we find

$$\rho(r) = -\frac{1}{4\pi r}\frac{d^2}{dr^2}e^{-\lambda r}$$

$$= \frac{-\lambda^2 e^{-\lambda r}}{4\pi r} \quad \text{for } r \neq 0.$$

For $r \to 0$,

$$V = \frac{e^{-\lambda r}}{r} \to \frac{1}{r}.$$

The relation

$$\nabla^2 \frac{1}{r} = -4\pi\delta(r)$$

was proved in problem (II-1). Therefore we have

$$\rho(r) = \delta(r) - \frac{\lambda^2}{4\pi}\frac{e^{-\lambda r}}{r}.$$

As $r \to 0$, only the term $\delta(r)$ contributes to the total charge since the volume is proportional to r^3.

(III-3) Let E, E', and E'' stand for the electric field vectors of the incoming wave, the refractive wave, and the reflected wave, respectively. The linear momentum density of the plane wave is known to be (e.g., see p. 200, Classical Electrodynamics by J. D. Jackson):

$$G_1 = \frac{\mu\epsilon}{4\pi c}E'H' = \frac{n^2}{4\pi c}E'H' \approx \frac{\epsilon}{4\pi c}E'H' \quad \text{with } H = \sqrt{\epsilon}\,E' = nE'$$

for the refracted wave in the dielectric medium, and

$$G_2 = \frac{EH}{4\pi c}$$

for the incoming wave in the air, and

$$G_3 = \frac{E''H''}{4\pi c}$$

for the reflected wave in the air. The momentum per unit time per unit area carried to the medium by the incoming plane wave is $P_2 = cG_2$, and that carried away by the refracted and reflected waves is $P_{1,3} = \frac{c}{n}G_1 - cG_3$. Using the second law of motion, we find that the pressure on the surface of the medium should equal the difference between P_2 and $P_{1,3}$, i.e.,

$$\text{Pressure} = P_2 - P_{1,3} = \frac{1}{4\pi}(EH - nE'H' + E''H'').$$

Now using the simple relations between E, E' and E'' for the case of normal incidence

$$E' = \frac{2}{n+1}E$$

and

$$E'' = \frac{n-1}{n+1}E \quad \text{with} \quad H' = nE', \quad E'' = H'', \quad E = H,$$

we find

$$\text{Pressure} = \frac{1}{4\pi}E^2[1 + (\frac{n-1}{n+1})^2 - (\frac{2n}{n+1})^2]$$

$$= \frac{1}{12.56} \times (10^{12})[2 - \frac{4n^2 + 4n}{(n+1)^2}]$$

$$= -1.4 \times 10^{10} \frac{\text{dyne}}{\text{cm}^2}.$$

(III-4)

a. Let λ be the charge per unit length on the cylinder with radius r_1. From Gauss's law we have

$$\iint E_r dA = \frac{\lambda}{\epsilon} \qquad (1)$$

where the surface S is defined to be a cylinder of radius r and of unit length. Since E_r is a function of r only, (1) immediately leads to

$$E_r = \frac{\lambda}{2\pi\varepsilon r}. \tag{2}$$

The potential difference between the two cylinders is

$$V = \int_{r_1}^{r_2} E_r \, dr = \frac{\lambda}{2\pi\varepsilon_o}\left(\frac{1}{K}\ell n\frac{r_3}{r_1} + \ell n\frac{r_2}{r_3}\right) \tag{3}$$

from which we find the capacitance C

$$C = \frac{\lambda L}{V} = \frac{2\pi\varepsilon_o L}{\frac{1}{K}\ell n\frac{r_3}{r_1} + \ell n\frac{r_2}{r_3}} \tag{4}$$

b. From (3), we can solve for the charge density

$$\lambda = \frac{2\pi\varepsilon_o V}{\frac{1}{K}\ell n\frac{r_3}{r_1} + \ell n\frac{r_2}{r_3}}. \tag{5}$$

From Gauss's law we get

$$\begin{aligned} E &= \frac{\lambda}{2\pi\varepsilon_o Kr} \quad \text{for } r_1 < r < r_3 \\ &= \frac{\lambda}{2\pi\varepsilon_o r} \quad \text{for } r_3 < r < r_2 \end{aligned} \tag{6}$$

from which we obtain the displacement in the dielectric medium

$$D = \frac{\lambda}{2\pi r} \quad r_1 < r < r_2 \tag{7}$$

and the polarization

$$\begin{aligned} P \equiv D - \varepsilon_o E &= \frac{(K-1)\lambda}{2\pi Kr} \quad r_1 < r < r_3 \\ &= 0 \quad r_3 < r < r_2. \end{aligned} \tag{8}$$

c. When the potential difference between the two cylinders is kept constant, the system can no longer be isolated. There has to be some source of charge (battery) to supply energy. Let C' be the capacitance of the system without the dielectric material. From (4) we find

$$C' = \frac{2\pi\varepsilon_o L}{\ln \frac{r_2}{r_1}} . \tag{9}$$

The work needed is

$$\text{work needed} = \frac{1}{2}C'V^2 - \frac{1}{2}CV^2 - (Q' - Q)V$$

where Q' is the final total charge on one cylinder. Using $Q' = C'V$, we get

$$\text{Work needed} = \frac{V^2}{2}(C - C')$$

$$= \frac{V^2}{2} 2\pi\varepsilon_o L \left(\frac{1}{\frac{1}{K}\ln \frac{r_3}{r_1} + \ln \frac{r_2}{r_3}} - \frac{1}{\ln \frac{r_2}{r_1}} \right)$$

where we have used (4) and (9).

(III-5)

a. The magnetic induction for such a toroid is

$$B = \mu\mu_o \frac{NI}{2\pi d} .$$

Therefore the total flux passing through the cross section of the toroid is

$$\varphi = \mu\mu_o \frac{NI}{2\pi d} A \quad \text{for} \quad d \gg \sqrt{A} .$$

b. The equivalent circuit equation for magnetic flux is

$$\varphi_1 R_1 + \varphi_2 R_2 = NI$$

where R_1 is the reluctance of the path in the iron ring and R_2 is the reluctance of the air gap. Using

$$\varphi_1 = \varphi_2 = \varphi, \quad R_1 = \frac{(2\pi d - \delta)}{\mu\mu_o A}, \quad \text{and } R_2 = \frac{\delta}{\mu_o A}$$

we obtain

$$\varphi = \mu_o \frac{NIA}{\frac{(2\pi d - \delta)}{\mu} + \delta} = \mu\mu_o \frac{NIA}{2\pi d + (\mu - 1)\delta}$$

$$\approx \mu\mu_o \frac{NIA}{(\mu\delta + 2\pi d)}$$

c. Field energy in iron $= \frac{1}{2\mu\mu_o} B^2 \times (\text{volume}) = \dfrac{\mu\mu_o N^2 I^2 A d}{4\pi(\frac{\mu\delta}{2\pi} + d)^2}$

Field energy in the gap $= \dfrac{\mu^2 \mu_o N^2 I^2 A\delta}{8\pi^2 (\frac{\mu\delta}{2\pi} + d)^2}$.

For $\mu >> 1$, we see that it takes comparable amount of energy to fill up the gap even for very small δ.

d. The self-inductance of each turn equals $d\varphi/dI$. There are N turns and all of the flux links with each turn; therefore it follows that

$$L = N\frac{d\varphi}{dI}.$$

Using

$$\varphi = \frac{\mu\mu_o NAI}{\mu\delta + 2\pi d}$$

we find

$$L = \frac{N^2 \mu\mu_o A}{\mu\delta + 2\pi d}.$$

(III-6)

a. The magnetic induction due to the large loop at the center of the ring is

$$B = \frac{\mu_0 I}{2b}. \tag{1}$$

The total flux is

$$\varphi = \vec{B} \cdot \vec{A}$$

$$= \frac{\pi a^2 \mu_0 I}{2b} \cos\theta \quad \text{for} \quad a \ll b \tag{2}$$

where θ is the angle between the two loops. If the small loop is rotating with constant angular velocity ω, we have $\theta = \omega t$. The induced electromotive force is

$$\text{emf} = -\frac{d\varphi}{dt}.$$

From (2) we get

$$\text{emf} = \frac{\pi a^2 \mu_0 I}{2b}\,\omega \sin\omega t. \tag{3}$$

According to Ohm's law, the current in the small loop, I_1, is

$$I_1 = \left(\frac{\pi a^2 \mu_0 I}{2bR}\right)\omega \sin\omega t \tag{4}$$

b. The input mechanical power is $\tau \frac{d\theta}{dt}$ which should be equal to the induced electrical energy $I_1 \frac{d\varphi}{dt}$ in order to conserve energy. Using (3) and (2) we obtain

$$\tau = I_1 \frac{d\varphi}{d\theta} = \left(\frac{\pi a^2 \mu_0 I}{2bR}\right)(\sin\omega t)(\omega \frac{d\varphi}{d\theta})$$

$$= \left(\frac{\pi a^2 \mu_0 I}{2b}\right)^2 \frac{\omega \sin^2 \omega t}{R} \tag{5}$$

c. From (2) we find that the mutual inductance of the two loops is

$$M = \frac{d\varphi}{dI} = \frac{\pi a^2 \mu_0}{2b} \cos\theta = \frac{\pi a^2 \mu_0}{2b} \cos\omega t$$

The induced emf in the large loop therefore is

$$E = -M\frac{dI_1}{dt} = -M\frac{d}{dt}\left(\frac{\pi\omega a^2 \mu_0 I \sin\omega t}{2bR}\right)$$

$$= -\frac{\pi^2 I\omega^2 a^4 \mu_0^2}{4b^2 R} \cos^2 \omega t$$

$$= -\left(\frac{\pi a^2 \mu_0 \omega \cos\omega t}{2b}\right)^2 \frac{I}{R}.$$

(IV-1)

$$E = \frac{q^2}{a} - \frac{q^2}{a} - \frac{q^2}{a} = -\frac{q^2}{a}.$$

(IV-2)

$$\frac{\text{electrical}}{\text{gravitational}} = \frac{(4.8 \times 10^{-10})^2}{10^{-27} \times 10^{-27} \times 2000 \times 6.67 \times 10^{-8}} \sim 10^{39}$$

where we have used $e = 4.8 \times 10^{-10}$ esu, $m_P \sim 2000\ m_e$, $m_e \sim 10^{-27}$ gm, and $G = 6.67 \times 10^{-8}$ in cgs units.

(IV-3) The potential energy between a point charge q and a plane is just half of that of point charges q and -q separated by a distance 2d. Therefore we have

$$E = \frac{-q^2}{4d}.$$

(IV-4) According to Gauss' law we find the field due to the charge on the sheet is $\pm 2\pi\sigma$ for $x \gtrless 0$. Using the superposition principle, we obtain the resultant field

$$E = E_o \pm 2\pi\sigma \quad \text{for} \quad x \gtrless 0.$$

(IV-5) Let q' be the charge on the inner shell. The potential at the inner shell is the sum of the potentials due to q and q', respectively:

$$V = \frac{q}{r_2} + \frac{q'}{r_1}$$

which is zero since the inner shell is grounded. Therefore

$$q' = -\frac{r_1}{r_2} q.$$

(IV-6) Using

$$\oint \vec{B} \cdot d\vec{\ell} = \mu_o i = \mu_o \frac{r^2}{R^2} I$$

we get

$$B = \frac{\mu_o r}{2\pi R^2} I.$$

(IV-7) For direct current:

$$\frac{V_N}{V_{2N}} = \frac{R_N}{R_{2N}} = \frac{1}{2}.$$

For high-frequency current:

$$\frac{V_N}{V_{2N}} = \frac{Z_N}{Z_{2N}} \approx \frac{L_N}{L_{2N}} = \frac{N^2}{(2N)^2} = \frac{1}{4}$$

where L_i is the self-inductance of the toroid i.

(IV-8) The force between the wire and the slab is the same as that between two wires of current I at a distance 2d. Therefore

$$F = -2 \times 10^{-7} \frac{I^2}{2d} \frac{\text{Newton}}{\text{meter}} \quad \text{(Attractive).}$$

(IV-9) The magnetic induction B at the center of the loop is

$$B = \frac{\mu_o I}{2r}$$

from which we find the torque is

$$\text{Torque} = HMV = \frac{BMV}{\mu_o}$$

$$= \frac{VMI}{2r}.$$

Here we have used the convention of permanent magnets. If one uses the convention of current loops, M should be replaced by $M' \equiv (M/\mu_o)$ and H by B; similarly for problems (VI-2) and (VII-3).

(IV-10) The induced emf is

$$\text{emf} = -\frac{d\varphi}{dt} = BLv$$

$$= (2 \times 10^{-5}) \times 1 \times \left(\frac{18 \times 10^4}{3.6 \times 10^3}\right) = \left(2 \times 5 \times 10^{-4} \frac{\text{wb}}{\text{sec}}\right)$$

$$= 10^{-3} \text{ volt.}$$

(IV-11) The work needed is equal to the potential energy difference which is $2mE \cos \alpha$.

(IV-12) Since the tangential component of the electric intensity is continuous across the boundary surface, we have $E = E_o$. It follows then that $D = KE = KE_o$.

(IV-13) $R = 2\sqrt{L/C}$ as we proved in problem (I-1).

(IV-14) Since

$$\text{intensity} = \left\langle \frac{c}{2}\left(\epsilon_o E^2 + \frac{B^2}{\mu_o}\right)\right\rangle = \frac{c\epsilon_o}{2} E_o^2$$

$$= \frac{c}{2\mu_o} B_o^2$$

where E_o and B_o are the peak values, we have

$$E_o = \sqrt{\frac{2 \times 1.3 \times 10^3}{8.85 \times 10^{-12} \times 3 \times 10^8}} \approx 10^3 \frac{\text{volts}}{\text{m}}$$

and

$$B_o = \frac{E_o}{c} \simeq \frac{10^3}{3 \times 10^8} \simeq 3 \times 10^{-6} \frac{\text{wb}}{\text{m}^2}.$$

(IV-15) Using Gauss's law we find

$$\epsilon_1 E_{1n} - \epsilon_2 E_{2n} = \frac{\delta}{\epsilon_o}.$$

The tangential component of the E-field is continuous. Therefore $E_{1t} = E_{2t}$.

(IV-16)

$$\frac{\vec{P} \cdot \vec{r}}{4\pi\epsilon_o r^3} \propto \frac{1}{r^2}$$

(IV-17) 0.5 Gauss.

(IV-18) The electric potential inside a conducting shell is

$$\frac{Q}{4\pi\epsilon_o R}.$$

(IV-19)

$$\text{Energy density} = \frac{1}{2}(\epsilon_o E^2 + \frac{1}{\mu_0} B^2).$$

$$\text{Momentum density} = \frac{\vec{E} \times \vec{H}}{c^2} \text{ in mks units.}$$

(IV-20)

$$\text{Poynting vector} = \vec{E} \times \vec{H} = (RI\hat{k}) \times \frac{I}{2\pi b}\hat{\theta}$$

$$= -\frac{RI^2}{2\pi b}\hat{r}_1 \quad \text{(inward).}$$

(V-1) The reactance is

$$\frac{1}{Z} = \frac{1}{i\omega L} + \frac{1}{1/i\omega C} = \frac{1 - \omega^2 CL}{i\omega L}$$

or

$$Z = \frac{i\omega L}{1 - \omega^2 CL}.$$

Since

$$\omega^2 < \omega_0^2 \equiv \frac{1}{LC}$$

we have

$$Z = \frac{i\omega L}{1 - \omega^2/\omega_0^2}$$

which is inductive.

(V-2) 0.

(V-3) The final capacitance is $\frac{C}{N}$, the potential difference is

$$\frac{Q}{C/N} = N\frac{Q}{C} = NV \quad \text{since} \quad V = \frac{Q}{C} = \frac{\text{charge on each capacitor}}{\text{capacitance of each capacitor}}.$$

(V-4) 4L.

(V-5) Using the fact that the displacement D is not affected by the existence of the dielectric medium, we find

$$W = \frac{D^2}{2K\epsilon_o} = \frac{D_o^2}{2K\epsilon_o} = \frac{W_o}{K},$$

or equivalently

$$W = \frac{1}{2}CV^2 = \frac{1}{2}(KC_o)\left(\frac{V_o}{K}\right)^2 = \frac{W_o}{K}.$$

(V-6) According to the inverse square law we have

$$E \propto \frac{1}{r}.$$

(V-7) Since the tangential component of the E-field is continuous, we have

$$\vec{E} + \vec{E}_R = 0 \text{ or } \vec{E}_R = -\vec{E},$$

where $\vec{E}_R$ is the electric vector in the reflected wave. Therefore $\vec{E}_R$ is shifted 180° in phase with respect to $\vec{E}$.

(V-8) Greater. (Metal spheres allow redistribution of charge in response to the attractive force, so that the effective separation distance is decreased.)

(V-9)

$$F = -\frac{\mu_o I^2}{2\pi d}.$$

The force is attractive.

(V-10)

a. The line integral of the magnetic intensity around any closed curve is equal to the the total magnetizing current through the surface bounded by the curve, i.e.,

$$\oint H d\ell = I = ni$$

where n is the number of turns of wire. Since the air gap is much smaller than the pole diameter, the induction in the air gap is approximately uniform. The above formula becomes

$$H_1\ell_1 + H_2\ell_2 = ni$$

where 1 and 2 refer to air and iron, respectively. H_1 and H_2 are related to B_1 and B_2 by

$$H_1 = \frac{B_1}{\mu_1} = \frac{B_1}{\mu_o} \quad \text{and} \quad H_2 = \frac{B_2}{\mu_2} = \frac{B_2}{\mu\mu_o}.$$

For very narrow air gap $B_1 \approx B_2$, we have

$$\frac{B}{i} = \frac{n\mu_o}{\ell_1 + \frac{\ell_2}{\mu}} \simeq 2.5 \frac{\text{webers}}{\text{m}^2\text{-amp.}}$$

b. For $i = 0.5$ amp, $\mu_o H = 1.25 \times 10^{-3}$ webers/m^2. When i increases to 1 amp, $\mu_o H = 2.5 \times 10^{-3}$. From the given B-H curve we find $B \sim 1.7$ webers/m^2 instead of 2.5 webers/m^2.

(V-11)

a. The system reaches the steady state when the induced emf cancels the potential difference of the battery, i.e., when

$$\epsilon = Bv\ell$$

$\Phi = B.\ell.x \quad \mathcal{E} = \frac{d\Phi}{dt}$

or

$\frac{dx}{dt} = V$

$$v = \frac{\epsilon}{B\ell}.$$

b. The current flowing in the bar at time t is

$$i = \frac{\varepsilon - Bv\ell}{R}$$

from which we find that the equation of motion for the bar is

$$m\dot{v} = F_x = \left(\frac{\varepsilon - Bv\ell}{R}\right)B\ell .$$

The solution is

$$v = \frac{\varepsilon}{B\ell}\left(1 - \exp\left(- \frac{B^2\ell^2}{mR} t\right)\right)$$

where we have used the condition that $v = 0$ at $t = 0$.

c. The equation of motion becomes

$$m\dot{v} = \left(\frac{\varepsilon - Bv\ell}{R}\right)B\ell - F .$$

Using the steady-state condition $\dot{v} = 0$ we obtain

$$v = \frac{\varepsilon}{B\ell} - \frac{FR}{(B\ell)^2} .$$

d. The current i under the conditions of part c is

$$i = \frac{\varepsilon - Bv\ell}{R} = \frac{F}{B\ell} .$$

The energy supplied by the battery is εi of which a power of Fv is being converted into mechanical work. The efficiency is

$$\text{Eff} = \frac{Fv}{\varepsilon i} = \frac{F\left\{\frac{\varepsilon}{B\ell} - \frac{FR}{(B\ell)^2}\right\}}{\frac{F\varepsilon}{B\ell}} = 1 - \frac{FR}{\varepsilon B\ell}$$

(V-12)

a. The circuit equation is

$$\frac{Q}{C} + \frac{dQ}{dt} R = 0$$

from which we find

$$Q = Q_o e^{-t/RC}.$$

b. Since the charges are uniformly distributed, the total charge outside the ring of radius ρ is

$$q = \frac{Q}{r^2}(r^2 - \rho^2) = \frac{Q_o(r^2 - \rho^2)}{r^2} e^{-t/RC},$$

from which we find that the current flowing into the ring of radius ρ is

$$i = -\frac{dq}{dt} = \frac{Q_o(r^2 - \rho^2)}{r^2 RC} e^{-t/RC}.$$

c. Using Ampere's law, we find that the magnetic field is

$$B_\theta = \frac{\mu_o I}{2\pi r} = \frac{\mu_o Q_o}{2\pi r RC} e^{-t/RC}.$$

d.

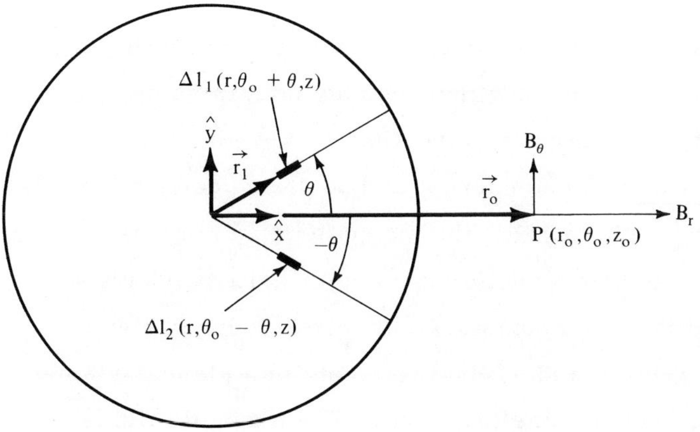

Figure 1 (projected view from the top)

The reason why there is only a θ component of the magnetic field is as follows (Fig. 1). To find the magnetic fields B_r and B_z at an arbitrary point $P(r_o, \theta_o, z_o)$, where r_o, θ_o, z_o refer to cylindrical coordinates, we first calculate the fields at that point due to two current segments $\Delta\ell_1$ and $\Delta\ell_2$. We define $\Delta\ell_1$ to be at $(r, \theta_o + \theta, z)$ and $\Delta\ell_2$ to be at $(r, \theta_o - \theta, z)$. The current directions are along the radius vectors. The fields due to $\Delta\ell_1$ and $\Delta\ell_2$ are: (For simplicity, we have assumed P is in the x-z plane.)

$$\vec{B}_1 \propto \vec{i}_1 \times (\vec{r}_1 - \vec{r}_0)$$

$$\propto (\cos\theta\,\hat{x} + \sin\theta\,\hat{y}) \times [(r\cos\theta - r_0)\hat{x} + r\sin\theta\,\hat{y} + (z - z_0)\hat{z}]$$

and

$$\vec{B}_2 \propto (\cos\theta\,\hat{x} - \sin\theta\,\hat{y}) \times [(r\cos\theta - r_0)\hat{x} - r\sin\theta\,\hat{y} + (z - z_0)\hat{z}]$$

or

$$\vec{B}_1 + \vec{B}_2 \propto 2\cos\theta(z - z_0)(\hat{x} \times \hat{z}) = -2\cos\theta(z - z_0)\hat{y}$$

$$= -2\cos\theta(z - z_0)\hat{\theta},$$

from which we see B_r and B_z due to the two current segments vanish. If we integrate the field over r and θ and sum over the two plates, we get the fields due to the two plates. From the consideration above, we conclude that the only nonvanishing component of the field is along θ direction.

Alternative proof: Since current and position vectors are polar, it follows from Ampere's law that the magnetic field is an axial vector. Assuming electromagnetic laws are invariant under space reflection, we find that the magnetic field vector changes sign upon space reflection. Let P and Q be two points in a plane, symmetric with respect to the central wire (Fig. 2a). The magnetic fields at these two points are related to each other, as shown by the arrows in Fig. 2a, by rotational symmetry. Let P' be the reflected point of P in the mirror. The magnetic field at P' as shown in Fig. 2b is of opposite sign to the image of the field at P in the mirror. We then rotate Fig. 2b by 180° thus to get the situation in Fig. 2c. We find

P' coincides with Q but the field $B_r(B_z)$ at P' is of opposite sign to $B_r(B_z)$ at Q. It follows that B_r and B_z must be zero.

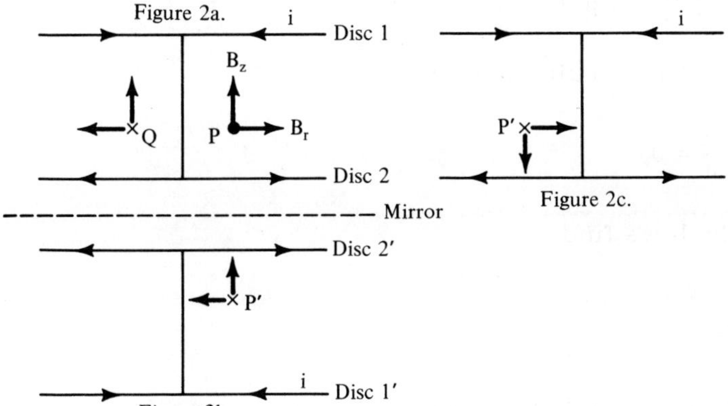

Figure 2a.
Figure 2b.
Figure 2c.

(V-13)

$$\text{Speed} = \text{frequency} \times \text{wavelength}$$

$$= 4 \times 10^{14} \times 5 \times 10^{-7}$$

$$= 2 \times 10^{8} \text{ m/sec.}$$

$$n = \frac{c}{v} = \frac{3}{2}$$

$$\lambda_o = n\lambda = 7.5 \times 10^{-7} \text{ meters.}$$

(VI-1)

a. The magnetic field outside the plates is zero and that between the plates is approximately constant. Using

$$\oint \vec{B} \cdot d\vec{s} = \mu_o I$$

where $d\vec{s}$ is perpendicular to the plane of the paper, we find

$$Bb \simeq \mu_o I \text{ or } B \simeq \frac{\mu_o I}{b} \text{ (into the plane of the paper).}$$

Here we have assumed $b >> a$ so that any term depending on a/b is negligible.

b. The self-inductance is

$$L = \left|\frac{d\varphi/dt}{dI/dt}\right| = \left|\frac{d\varphi}{dI}\right| = \frac{d(BA)}{dI} = \frac{d}{dI}\left(\frac{\mu_o I}{b}\ell a\right) = \frac{\mu_o \ell a}{b}.$$

c. The circuit equation is

$$V_o - L\frac{di}{dt} = 0,$$

from which we find

$$i = \frac{V_o}{L}t.$$

d. The self-inductance for two strips of length x is

$$L_1 = \frac{\mu_o xa}{b};$$

therefore

$$V_x = L_1\frac{di}{dt} = \frac{\mu_o xaV_o}{bL}.$$

e. The rate of energy flow is

$$\text{Energy} = V_x i = \left(\frac{\mu_o xaV_o}{bL}\right)\left(\frac{V_o}{L}t\right)$$

$$= \frac{\mu_o xa}{b}\left(\frac{V_o}{L}\right)^2 t.$$

(VI-2) With $L >> R$, the two loops act like two magnetic dipoles with magnetic moment $\mu_o IA$, where A is the area of the loop. The induction B at loop two due to loop one can be resolved into two components; B_r, in the direction of increasing r, and B_θ in the direction of increasing θ. We find

$$B_r = \frac{\mu_o}{2} IR^2 \frac{\cos\theta}{L^3} = \frac{\mu_o}{2} IR^2 \frac{x}{(x^2 + y^2)^2}$$

$$B_\theta = \frac{\mu_o}{4} IR^2 \frac{\sin\theta}{L^3} = \frac{\mu_o}{4} IR^2 \frac{y}{(x^2 + y^2)^2},$$

where $x = L\cos\theta$, $y = L\sin\theta$, and $\mu_o I\pi R^2$ is the equivalent dipole moment of each loop. The torque on loop two is

$$\vec{\tau} = I\vec{A} \times \vec{B} \quad (= \mu_o I\vec{A} \times H)$$

or

$$\tau = I\pi R^2 (B_r \sin\theta + B_\theta \cos\theta)$$

$$= \frac{3\pi\mu_o I^2 R^4 \sin\theta \cos\theta}{4L^3}$$

The direction of the torque is pointing into the plane of the paper. The force on loop two is

$$\vec{F} = \vec{F}_\theta + \vec{F}_r$$

where

$$F_\theta = IR^2 \frac{dB_\theta}{dx}$$

$$= -\frac{\mu_o}{4} I^2 R^4 y \frac{2 \times 2x}{(x^2 + Y^2)^3}$$

$$= -\mu_o (IR^2)^2 \frac{xy}{(x^2 + y^2)^3}$$

$$= -\mu_o (IR^2)^2 \frac{\sin\theta \cos\theta}{L^4}$$

and

$$F_r = IR^2 \frac{dB_r}{dx}$$

$$= \frac{1}{2}\mu_o (IR^2)^2 \frac{y^2 - 3x^2}{(x^2 + y^2)^3}$$

$$= \frac{1}{2}\mu_o (IR^2)^2 \frac{\sin^2 \theta - 3 \cos^2 \theta}{L^4}.$$

(VI-3) The field at P due to the wire other than the portion of the half-circle equals that due to an infinitely long wire with current I at a distance r. The field produced by the current along the half-circle is just half of that due to a full-circle with current I. Therefore

$$B = 2 \times \frac{1}{2}[\frac{\mu_o I}{2\pi r}] + \frac{1}{2}[\frac{\mu_o I}{2r}] = \frac{\mu_o I}{4\pi r}(2 + \pi)$$

the direction of which is perpendicular to the plane of the paper.

The above conclusion is obvious if we add another identical loop as shown in the following figure. Now we have two infinite long wires and a full-circle, both with current I. The fields due to the two wires simply add up at P. The field of each loop is one half of that of the whole circuit due to rotational symmetry.

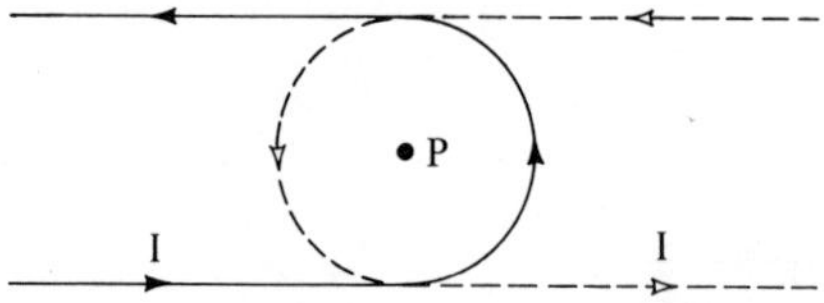

(VI-4) The wave equation in the cavity is

$$(\frac{\partial^2}{\partial x^2} + \frac{\partial^2}{\partial y^2} + \gamma^2)\psi = 0 \tag{1}$$

where

$$\gamma^2 = (\mu_o \epsilon_o \frac{\omega^2}{c^2} - k^2);$$

and ψ stands for either the E- or B-field. The boundary conditions of the fields are

$$\frac{\partial\psi}{\partial n} = 0 \quad \text{at} \quad x = 0, a \quad \text{and} \quad y = 0, b.$$

The general solution of (1) is

$$\psi_{mn}(x, y) = B_o \cos\frac{m\pi x}{a} \cos\frac{n\pi y}{b} \qquad m, n = 1, 2, 3 \ldots \tag{2}$$

Substituting (2) into (1) we find

$$\gamma_{mn}^2 = \pi^2\left(\frac{m^2}{a^2} + \frac{n^2}{b^2}\right). \tag{3}$$

The z-dependence of the fields is that appropriate to standing waves:

$$\psi_k(z) = A \sin kz + C \cos kz.$$

In order to satisfy the boundary conditions at $z = 0$ and $z = d$, k has to satisfy the following relation

$$k = \frac{\ell\pi}{d} \qquad \ell = 0, 1, 2 \ldots \tag{4}$$

Using (4), (3), and the definition of γ, we find that the cutoff frequency is

$$\omega_{mn1} = \frac{\pi c}{\sqrt{\mu_o\epsilon_o}}\left(\frac{m^2}{a^2} + \frac{n^2}{b^2} + \frac{\ell^2}{d^2}\right)^{1/2}.$$

For $a > b > d$, the lowest frequency corresponds to $m = 1$, $n = 0$, and $\ell = 0$. We find

$$\omega_{100} = \frac{\pi c}{a\sqrt{\mu_o\epsilon_o}}.$$

The corresponding field is

$$\psi_{100} = \psi_{10}(x, y)\psi_o(z) \sim \cos\frac{\pi x}{a}.$$

(VI-5)

a. Supplemented by Ohm's law $\vec{J} = \sigma\vec{E}$, Maxwell's equations become:

$$\nabla \times \vec{E} + \frac{\mu}{c}\frac{\partial \vec{H}}{\partial t} = 0$$

$$\nabla \times \vec{H} - \frac{4\pi\sigma}{c}\vec{E} = 0 \tag{1}$$

where we have neglected the term corresponding to the displacement current. For a harmonic wave with frequency ω, we can eliminate $\vec{E}$ from the above equations and obtain the following wave equation:

$$\nabla \times \nabla \times \vec{H} + \frac{4\pi\sigma\mu\omega}{c^2} i\vec{H} = 0 \tag{2}$$

where we have assumed that $\vec{H} = \vec{H}_o e^{i(\omega t - \vec{k} \cdot \vec{r})}$. For a plane wave, $\vec{H}$ is parallel to the surface. Therefore

$$\nabla \approx -\vec{n}\frac{\partial}{\partial z},$$

so that

$$\nabla \times \nabla \times \vec{H} = \nabla(\nabla \cdot \vec{H}) - \nabla^2\vec{H} = -\nabla^2\vec{H} = -\frac{\partial^2}{\partial z^2}\vec{H}.$$

Eq. (2) becomes

$$\frac{\partial^2}{\partial z^2}(\vec{n} \times \vec{H}) - \frac{4\pi\sigma\mu\omega}{c^2} i\vec{n} \times \vec{H} = 0, \tag{3}$$

where $\vec{n} \times \vec{H} = \vec{H}^t$.

b. Since the electric potential between two points is independent of path, we must have $E^t = E_c^t$, i.e., the tangential component of the E-field is continuous. For a perfect metal, H^t vanishes inside the metal since the surface charges move in response to the incoming plane wave. We have

$$\vec{H}^t = \frac{4\pi}{c}\vec{K} \text{ and } \vec{H}_c^t = 0 \tag{4}$$

where $\vec{K}$ is the surface current, occupying a layer of thickness of the order of the skin depth; $\vec{K} = \sigma\vec{E}^t$. For $z \to 0$, then equation (4) becomes $\vec{H}^t = \vec{H}_c^t$.

c. Using (3) we can solve for H_c^t

$$H_c^t = H^t e^{-\left(\frac{1+i}{\delta}\right)z} \tag{5}$$

where

$$\delta = \frac{c}{\sqrt{2\pi\mu\omega\sigma}}$$

is called the skin depth. From (5) and the second equation of (1), we find

$$\vec{E}^t = -\sqrt{\frac{\mu\omega}{8\pi\sigma}}\,(1+i)(\vec{n} \times \vec{H}^t)\Big|_{\text{at } z=0}$$

which leads to

$$Z = \sqrt{\frac{\mu\omega}{8\pi\sigma}}\,(1+i).$$

(VI-6)

a. At the surface of the wire, the B-field is

$$B_\theta = \frac{\mu_o I}{2\pi a}, \quad B_r = 0 \tag{1}$$

where a is the radius of the wire. The electric field is related to the potential difference V which equals IR;

$$E_L = \frac{V}{L} = \frac{IR}{L} = \frac{I}{\sigma A}. \tag{2}$$

The direction of the E-field is along the wire. The Poynting vector at the surface of the wire in vector notation is

$$\vec{S} = \frac{1}{\mu_o}\vec{E} \times \vec{B} = \frac{1}{\mu_o}\left(\frac{IR}{L}\right)\left(\frac{\mu_o I}{2\pi a}\right)(-\hat{r}_1)$$

$$= -\frac{I^2 R}{2\pi a L}\hat{r}_1 \quad \text{(inward)} \tag{3}$$

where $\hat{r}_1$ is the unit vector pointing along the radius direction. The rate of the field energy flowing into a wire segment of length L is

$$W = 2\pi a L S$$

$$= I^2 R$$

which is the Joule heat produced.

(VII-1)

a. According to Ampere's law, we have

$$\oint \vec{B} \cdot d\ell = \mu_o I' \tag{1}$$

where I' is current through the surface bounded by the closed curve over which the integral is to be carried out.

i. For $r < a$, $I' = \frac{Ir^2}{a^2}$.

From (1) we get

$$B_\theta = \frac{\mu_o I'}{2\pi r} = \frac{\mu_o I r}{2\pi a^2}. \tag{2}$$

ii. For $a < r < b$, $I' = I$. Equation (1) becomes

$$B_\theta = \frac{\mu_o I'}{2\pi r} = \frac{\mu_o I}{2\pi r}. \tag{3}$$

iii. For $b < r < c$,

$$I' = I - I\frac{r^2 - b^2}{c^2 - b^2} = I\left(\frac{c^2 - r^2}{c^2 - b^2}\right).$$

Equation (1) becomes

$$B_\theta = \frac{\mu_0 I}{2\pi r}\frac{(c^2 - r^2)}{(c^2 - b^2)}.$$

iv. For $r > c$, $I' = 0$. Therefore $B_\theta = 0$. The components along radial direction and z-axis are identical to zero for all four regions.

b. The self-inductance is defined as

$$L = \left|\frac{d\varphi/dt}{dI/dt}\right| = \left|\frac{d\varphi}{dI}\right| \tag{4}$$

where φ is the total flux passing through the region ABCD. Using (3) we find

$$\varphi = \ell\int_a^b dr\, B_\theta$$

$$= \frac{\ell\mu_0 I}{2\pi}\ \ell n\ \frac{b}{a}; \tag{5}$$

therefore

$$L = \frac{\ell\mu_0}{2\pi}\ \ell n\ \frac{b}{a}.$$

(VII-2)

a.

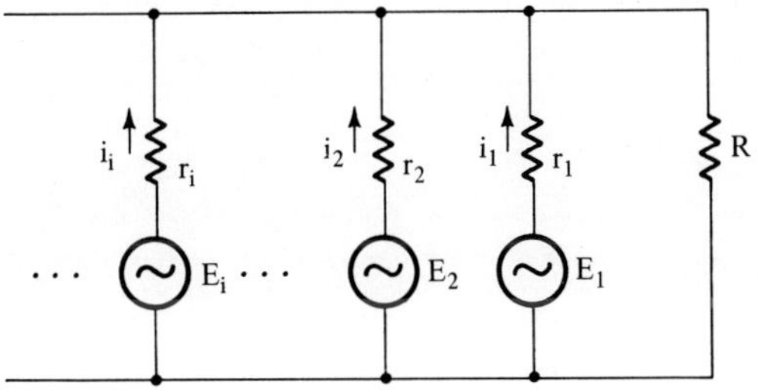

Let us first prove a general theorem that any such black box can be replaced by an effective emf and an effective resistor connected in series. The general arrangement of many emf's and resistors in the black box is shown above. Let i_i be the current flowing through the ith loop. The circuit equation for each loop is

$$E_i = \sum_{j=1}^{N} a_{ij} i_j + b_i RI \qquad i = 1, \ldots, N \tag{1}$$

together with the relation of conservation of currents,

$$\sum_{j=1}^{N} i_j = I \tag{2}$$

where b_i and a_{ij} are functions of the resistors. Solving for i_j in (1) we have

$$i_j = \sum_{i}^{N} (E_i - b_i RI) A_{ij} \qquad j = 1, \ldots, N \tag{3}$$

where (A_{ij}) is the inverse of (a_{ij}). Substituting (3) into (2) we have

$$\sum_{j,i}^{N} (E_i - b_i RI) A_{ij} = I$$

or

$$\sum_{j,i} E_i A_{ij} = I(1 + R \sum_{j,i}^{N} b_i A_{ij}). \tag{4}$$

Let

$$E = \frac{\sum_{j,i} E_i A_{ij}}{\sum_{j,i}^{N} b_i A_{ij}}$$

and

$$r = \frac{1}{\sum\limits_{j,i}^{N} b_i A_{ij}}$$

(4) becomes

$$E = I(R + r) \tag{5}$$

where E is the effective emf and r is the effective resistance. Using the given data we find $E = 12$ volts and $r = 2$ ohms. Therefore R should equal 118 ohms in order to get a current of 0.1 amp.

b.

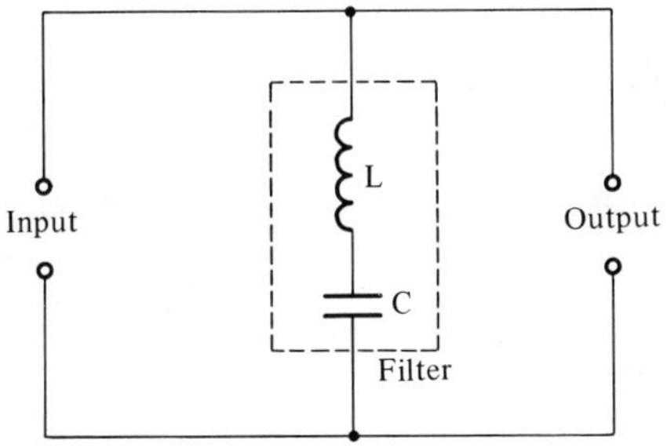

In order to attenuate 60 cps ripple voltage the filter must have very low impedance at that frequency. Therefore we have the relation

$$LC = \frac{1}{\omega^2} = \left(\frac{1}{120\pi}\right)^2 = 7 \times 10^{-6}.$$

(VII-3)

a.

$\overrightarrow{\vec{\mu}_1}$ $\overleftarrow{\vec{\mu}_2}$ $\quad \text{Energy}_{max} = -\mu_1 \cdot H_2 = -\mu_1\left(\frac{\mu_2 \cos\pi}{2\pi r^3 \mu_0}\right) = \frac{\mu_1\mu_2}{2\pi\mu_0 r^3}$

$\overrightarrow{\vec{\mu}_1}$ $\overrightarrow{\vec{\mu}_2}$ $\quad \text{Energy}_{min} = -\mu_1 \cdot H_2 = -\mu_1\left(\frac{\mu_2 \cos 0}{2\pi r^3 \mu_0}\right) = -\frac{\mu_1\mu_2}{2\pi\mu_0 r^3}.$

(VII-4) The voltage as a function of t is

$$V = V_o(1 - e^{-t/RC}).$$

Using V_o = 100 volts and V = 50 volts, we find

$$t = RC \ln 2 = (\ln 2) \times 10^{-4} \text{ seconds.}$$

(VII-5) The component of the velocity of the electron perpendicular to the field is

$$v_\perp = \frac{10^4}{\sqrt{2}} \frac{\text{cm}}{\text{sec}} = 7.071 \times 10^3 \frac{\text{cm}}{\text{sec}}.$$

From the equilibrium condition

$$\frac{mv_\perp^2}{R} = \frac{ev_\perp H}{c}$$

we find the radius R is

$$R = \frac{cv_\perp m}{eH} = \frac{(3 \times 10^{10})(7.1 \times 10^3)(0.91 \times 10^{-27})}{(4.80 \times 10^{-10})(10^4)} = 4.0 \times 10^{-8} \text{ cm.}$$

From $v_\perp$ and R we find the frequency is $f = v_\perp / 2\pi R = 2.8 \times 10^{10}$ hertz. Therefore the electron is travelling in a helix, with radius R and pitch $v_\perp (\frac{2\pi mc}{eH}) = 2\pi R = 25 \times 10^{-8}$ cm.

(VII-6) This system can be replaced by two wires, A and B, of radii a and b, respectively, as shown in the figure. Both wires carry the same current density J as that of the original wire. The current on the wire B is of opposite sense, while the current on the wire A is of the same sense as that in the original wire. According to the superposition principle, the field everywhere in space is the same as before. Using Ampere's law we find the induction B_1 due to the current on the wire A is

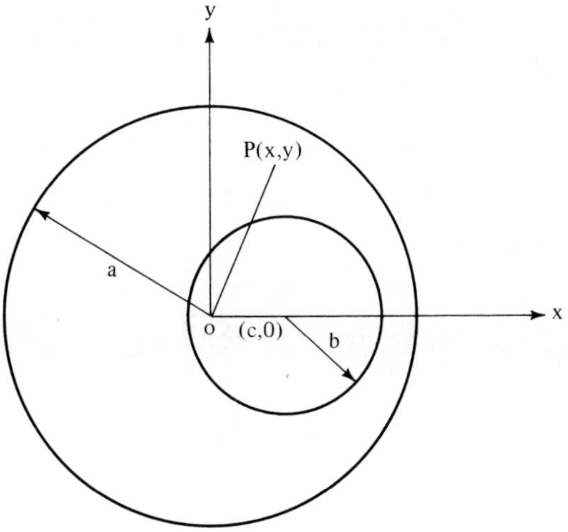

$$B_1 = \frac{\mu_o i}{2\pi r}$$

or

$$B_1 = \frac{\mu_o I r}{2\pi(a^2 - b^2)} \quad \text{for} \quad r < a \tag{1}$$

$$= \frac{\mu_o I a^2}{2\pi r(a^2 - b^2)} \quad \text{for} \quad r > a \tag{2}$$

where we have used the relation

$$J = \frac{I}{\pi(a^2 - b^2)}$$

and

$$i = J \times (\text{area})$$

$$= \frac{Ir^2}{(a^2 - b^2)} \quad \text{for} \quad r \leq a.$$

In the rectangular coordinate system, as shown in the figure, the vector induction $\vec{B}_1$ becomes

$$\vec{B}_1 = B_1 \frac{y}{\sqrt{x^2 + y^2}} \hat{x} - B_1 \frac{x}{\sqrt{x^2 + y^2}} \hat{y}. \tag{3}$$

We have assumed that the current in A is directed into the paper. Similarly, we can get the expression for B_2 produced by the current in the wire B

$$B_2 = \frac{\mu_o I r'}{2\pi(a^2 - b^2)} \quad \text{for} \quad r' \equiv \sqrt{(x - c)^2 + y^2} < b \tag{4}$$

$$= \frac{\mu_o I b^2}{2\pi r'(a^2 - b^2)} \quad \text{for} \quad r' > b \tag{5}$$

and the vector induction $\vec{B}_2$ is

$$\vec{B}_2 = - B_2 \frac{y}{\sqrt{(x - c)^2 + y^2}} \hat{x} + B_2 \frac{x - c}{\sqrt{(x - c)^2 + y^2}} \hat{y}. \tag{6}$$

The actual magnetic field $\vec{B}$ is the sum of $\vec{B}_1$ and $\vec{B}_2$:

$$\vec{B} = \vec{B}_1 + \vec{B}_2$$

$$= \left(\frac{B_1 y}{\sqrt{x^2 + y^2}} - \frac{B_2 y}{\sqrt{(x - c)^2 + y^2}} \right) \hat{x} - \left(\frac{B_1 x}{\sqrt{x^2 + y^2}} - \frac{B_2 (x - c)}{\sqrt{(x - c)^2 + y^2}} \right) \hat{y}.$$

In the region $r > a$, we find

$$\vec{B} = \frac{\mu_o I}{2\pi(a^2 - b^2)} \left(y\left(\frac{a^2}{x^2 + y^2} - \frac{b^2}{(x - c)^2 + y^2}\right)\hat{x} - \left(\frac{xa^2}{x^2 + y^2} - \frac{b^2 (x - c)}{(x - c)^2 + y^2}\right)\hat{y}.$$

In the region $r' < b$ (and $r < a$), we find

$$\vec{B} = \frac{\mu_o I}{2\pi(a^2 - b^2)} ((y - y)\hat{x} + ((x - c) - x)\hat{y})$$

$$= \frac{- \mu_o I c}{2\pi(a^2 - b^2)} \hat{y}.$$

In the region $r' > b$ but $r < a$, we find

$$\vec{B} = \frac{\mu_0 I}{2\pi(a^2 - b^2)}[(y - \frac{b^2 y}{(x - c)^2 + y^2})\hat{x} - (x - \frac{b^2(x - c)}{(x - c)^2 + y^2})\hat{y}].$$

(VIII-1) The effect due to the self-inductance of the loop is negligible since 10^{-4} sec is a long time here and the magnetic field due to the loop is negligible compared with the external field. Therefore we can take the current in the loop for $t \leq 10^{-4}$ sec as constant which equals 0.1 amp. The forces exerted on the sides of the loop are

$$F_h = IhB = 10^{-3} \text{ newtons} = 10^2 \text{ dynes} \tag{1}$$

and the forces on the top and the bottom of the loop are zero. The impulse is then

$$\hat{F} = F_h \Delta t = 10^2 \text{ dyne} \times 10^{-4} \text{ sec}$$

$$= 10^{-2} \text{ dyne-sec.} \tag{2}$$

The impulse torque is

$$\hat{\tau} = 0.02 \text{ dyne-cm-sec.} \tag{3}$$

The moment of inertia of the disc is

$$I = \int_0^{D/2} \rho h x^2 2\pi x dx = \frac{\pi}{2} \text{ gm-cm}^2. \tag{4}$$

The initial angular velocity is

$$\dot{\theta}_o = \frac{\hat{\tau}}{I} = \frac{0.02}{\pi/2} \sim \frac{0.013}{\text{sec}}.$$

When the wire is rotating with angular velocity $\dot{\theta}$, the induced emf is $\dot{\theta}BhD \cos \theta$. The induced current i is $\dot{\theta}BhD \cos \theta/R$. The torque exerted on the wire by the field is

$$\tau_1 = -DhiB\dot{\theta} = -\frac{B^2h^2D^2 \cos\theta}{R}\dot{\theta}$$

$$= -4 \times 10^{-8}\dot{\theta} \cos\theta \text{ newtons-m}$$

$$= -0.4\dot{\theta} \cos\theta \text{ dyne-cm.}$$

Let ω_o be the angular velocity corresponding to the natural frequency of the pendulum,

$$\omega_o = \frac{2\pi}{T} = 0.628/\text{sec.}$$

Since $\dot{\theta}_o \ll \omega_o$, we expect the amplitude of oscillation $A \ll 1$. The equation of motion of the disc is

$$I\ddot{\theta} + I\omega_o^2\theta = \tau_1$$

or

$$\ddot{\theta} + \omega_o^2\theta = \frac{\tau_1}{I} = -\frac{B^2h^2D^2 \cos\theta}{IR}\dot{\theta} \approx -\frac{B^2h^2D^2}{IR}\dot{\theta} = -b\dot{\theta} \tag{5}$$

where $b = \frac{B^2h^2D^2}{IR} = 0.26$ and we have assumed the small angle approximation $\cos\theta \approx 1$. The solutions of (5) are of the forms:

$$\theta = Ae^{\omega_+ t} + Be^{\omega_- t}$$

where

$$\omega_\pm = \frac{b \pm i\sqrt{4\omega_o^2 - b^2}}{2} = 0.13 \pm 0.615i.$$

Using the initial conditions $\theta = 0$ and $\dot{\theta} = \dot{\theta}_o$ at $t = 0$, we find

$$B = -A$$

and

$$A(\omega_+ - \omega_-) = \dot{\theta}_o$$

or

$$A = \frac{\dot{\theta}_o}{(\omega_+ - \omega_-)} = \frac{0.013}{1.23i} = -0.0106i \text{ (radians)}.$$

Therefore

$$\theta = -0.0106[e^{(-0.13+0.615i)t} - e^{(-0.13-0.615i)t}]i$$

$$= 0.0212e^{-0.13t} \sin(0.615t)$$

which shows the motion of a damped oscillator with amplitude $0.0212e^{-0.13t}$.

(VIII-2) The fact that the system has a resonant frequency at 1000 cycles/sec suggests that L and C are in series. The condition of resonance is

$$LC = \frac{1}{\omega_R^2}$$

$$= \frac{1}{(2\pi)^2 \times 10^6}. \tag{1}$$

From the condition that 100 volts dc gives a current of 0.1 amp we get

$$R = \frac{V}{I} = 1000 \text{ ohms}$$

which must be parallel to L and C. The total impedance is

$$\frac{1}{Z} = \frac{1}{R} + \frac{1}{j\omega L - j\frac{1}{\omega C}},$$

where $j = \sqrt{-1}$, or

$$Z = \frac{1}{\sqrt{\frac{1}{R^2} + \left(\frac{\omega C}{\omega^2 LC - 1}\right)^2}} e^{j\varphi} \tag{2}$$

where

$$\varphi = \tan^{-1} \frac{\omega CR}{(\omega^2 LC - 1)} .$$

For $V_{rms} = 100$ volts and $\omega = 2\pi \times 60$, it is known that $I_{rms} = 1$ amp. Therefore

$$|Z| = \frac{V_{rms}}{I_{rms}} = 100 \text{ ohms.} \qquad (3)$$

From (1), (2), and (3) we can solve for L and C. We find

$$\frac{1}{R^2} + \left(\frac{\omega C}{(\frac{\omega}{\omega_R})^2 - 1} \right)^2 = 10^{-4}$$

or

$$C = 2.3 \times 10^{-5} \text{ farads and } L = 1.1 \times 10^{-3} \text{ henries.}$$

(VIII-3) Maxwell's equations in differential form are

$$\nabla \cdot \epsilon \vec{E} = \rho = 0 \qquad (1)$$

$$\nabla \cdot \mu \vec{H} = 0 \qquad (2)$$

$$\nabla \times \vec{E} = - \frac{\mu \dot{\vec{H}}}{c} = \frac{i\omega\mu}{c} \vec{H} \qquad (3)$$

$$\nabla \times \vec{H} = \frac{\vec{J}}{c} + \frac{\epsilon \dot{\vec{E}}}{c}$$

$$= \frac{\sigma \vec{E}}{c} - \frac{i\omega\epsilon \vec{E}}{c} = (\frac{\sigma - i\omega\epsilon}{c}) \vec{E} \qquad (4)$$

where we have assumed the time dependence of fields is $e^{-i\omega t}$. Eliminating $\vec{H}$ or $\vec{E}$ from (3) and (4) we obtain

$$\nabla \times \nabla \times \vec{E} = \frac{i\omega\mu}{c^2} (\sigma - i\omega\epsilon) \vec{E} \qquad (5)$$

and

$$\nabla \times \nabla \times \vec{H} = (\sigma - i\omega\varepsilon)\frac{i\omega\mu}{c^2}\vec{H}. \tag{6}$$

For a plane wave propagating in the x-direction, $\vec{E}$ and $\vec{H}$ are functions of x and t only. Therefore we have for (5) and (6)

$$-\frac{d^2E_y}{d^2x} = \frac{\omega\mu}{c^2}(i\sigma + \omega\varepsilon)E_y$$

$$-\frac{d^2E_z}{d^2x} = \frac{\omega\mu}{c^2}(i\sigma + \omega\varepsilon)E_z$$

$$-\frac{d^2H_y}{d^2x} = \frac{\omega\mu}{c^2}(i\sigma + \omega\varepsilon)H_y$$

$$-\frac{d^2H_z}{d^2x} = \frac{\omega\mu}{c^2}(i\sigma + \omega\varepsilon)H_z.$$

The solutions to the above equations are of the form

$$A = A_o \exp(ikx)$$

where

$$k = \frac{\omega}{c}\sqrt{\mu\varepsilon + i\frac{\mu\sigma}{\omega}}$$

is the complex propagation vector.

(VIII-4) The kinetic energy of the incoming particle is minimum when all the particles in the final state are moving with the same velocity. This is because, in this case, all the final particles are at rest in the center of mass system. Therefore the invariant mass of the final particles is just the sum of the masses of all the final particles:

$$M = m_p + m_n + 2m_p + m_{\bar{p}} \sim 5\, m_p. \qquad (1)$$

The conservation of energy and momentum can be written as a 4-vector equation:

$$P_p + P_d = P_f \equiv P_{p_1} + P_n + P_{p_2} + P_{p_3} + P_{\bar{p}}. \qquad (2)$$

The squares of the 4-vector momenta are the invariants

$$P_p \cdot P_p = -m_p^2, \qquad P_d \cdot P_d = -m_d^2$$

and

$$P_f \cdot P_f = -M^2 \qquad (3)$$

where we have used the convention that $c = 1$. Squaring (2) and using (3) to simplify the result, we obtain

$$m_p^2 + m_d^2 + 2E_p m_d = M^2$$

where E_p is the total energy of the proton. We get

$$E_p = \frac{M^2 - m_p^2 - m_d^2}{2m_d} = 5m_p$$

or the kinetic energy of the incoming proton

$$K_p = E_p - m_p = 3752 \text{ MeV}.$$

(VIII-5) The circuit equation for the system is

$$L\frac{di}{dt} + iR = V$$

the solution of which is

$$i = \frac{V}{R}(1 - e^{-(R/L)t})$$

where we have taken the initial condition to be $i = 0$ at $t = 0$. The steady current is just V/R. The half time, t, at which the battery is delivering one-half its steady current, is obtained from the relation

$$\frac{1}{2} = 1 - e^{-(R/L)t}.$$

Therefore

$$t = \frac{L}{R} \ln 2.$$

(VIII-6) From the relation $\nabla \times \vec{A} = \vec{B}$, or

$$\frac{\partial A_z}{\partial y} - \frac{\partial A_y}{\partial z} = ky$$

$$\frac{\partial A_x}{\partial z} - \frac{\partial A_z}{\partial x} = kx$$

$$\frac{\partial A_y}{\partial x} - \frac{\partial A_x}{\partial y} = 0$$

we get two possible solutions for the vector potential

$$\vec{A} = (kxz, -kyz, 0) \text{ and } \vec{A}' = \frac{k}{2}(-x^2 + y^2)\hat{k}$$

which can be related through the Gauge transformation

$$\vec{A} = \vec{A}' + \frac{k}{2}\nabla(x^2 z - y^2 z);$$ therefore they are identical.

From the equation

$$-\nabla\varphi = \vec{B}$$

or $-\frac{\partial\varphi}{\partial x} = ky$, $-\frac{\partial\varphi}{\partial y} = kx$, and $-\frac{\partial\varphi}{\partial z} = 0$

we get the solution for the scalar potential

$$\varphi = -kxy + C,$$

where C is a constant.

(IX-1) The induction B at the edge is

$$B = \frac{\mu_0 I}{2\pi r} = 8 \times 10^{-3} \, \frac{\text{webers}}{\text{meter}}.$$

The force on an electron at the edge is

$$\vec{F} = e\vec{v} \times \vec{B}$$

or

$$F = 1.6 \times 10^{-19} \times 10^{3} \times 8 \times 10^{-3}$$

$$= 1.28 \times 10^{-18} \text{ newton (inward).}$$

Therefore the beam tends to shrink.

(IX-2) Inside the space filled with electrons, Maxwell's equations take the form

$$\nabla \times \vec{E} = i\frac{\omega}{c}\vec{B}; \qquad \nabla \cdot \vec{B} = 0$$

$$\nabla \times \vec{B} = -i\mu\epsilon\frac{\omega}{c}\vec{E} + \frac{4\pi\sigma\mu}{c}\vec{E}$$

and

$$\nabla \cdot \vec{E} = 0. \tag{1}$$

Eliminating either $\vec{B}$ or $\vec{E}$ with the first and third equations in (1), we get the wave equation

$$\left(\nabla^2 + \mu\epsilon\frac{\omega^2}{c^2} + \frac{i4\pi\sigma\mu\omega}{c^2}\right)\begin{Bmatrix}\vec{E}\\ \vec{B}\end{Bmatrix} = 0 \tag{2}$$

or

$$-k^2 + \mu\epsilon\frac{\omega^2}{c^2} - \frac{4\pi\mu Ne^2}{c^2 m} = 0 \tag{3}$$

where we have assumed the form $e^{-i\omega t+ikz}$ for $\vec{E}$ and $\vec{B}$. For $\mu = 1$ and $\epsilon = 1$, (3) takes the form

$$k^2 = \frac{\omega^2}{c^2}\left(1 - \frac{\omega_p^2}{\omega^2}\right) \qquad (4)$$

with

$$\omega_p^2 = \frac{4\pi Ne^2}{m},$$

where ω_p is called the plasma frequency. The index of refraction is defined by

$$n \equiv \frac{ck}{\omega}. \qquad (5)$$

From (4) and (5) we obtain

$$n = \sqrt{1 - \frac{\omega_p^2}{\omega^2}}.$$

For high-frequency $\omega > \omega_p$, n is real and the waves propagate freely through the plasma. For frequencies lower than the plasma frequency ω_p, n is pure imaginary. The waves are reflected from the upper atmosphere.

(IX-3) Let us first find the equivalent capacitance, C, of the system. Since C_2, C_3 are in series, we have

$$\frac{1}{C_{23}} = \frac{1}{C_2} + \frac{1}{C_3}$$

where C_{23} is the equivalent capacitor for C_2 and C_3, and since C_{23} and C_1 are parallel, we find

$$C = C_{23} + C_1$$

$$= \frac{C_2C_3}{C_2 + C_3} + C_1.$$

The electrical energy stored in the network is

$$W = \frac{1}{2}CV^2$$

$$= \frac{1}{2}(C_1 + \frac{C_2C_3}{C_2 + C_3})V^2$$

$$= \frac{(C_1C_2 + C_2C_3 + C_1C_3)V^2}{2(C_2 + C_3)}.$$

(IX-4) 10^{-5} mm of Hg.

(IX-5)

a. For $0 \leq t \leq t_1$, the velocity of the loop is just that of a free falling body, i.e., $v = gt$. The velocity at $t = t_1$ is

$$v_1 = gt_1 = \sqrt{2hg}.$$

b. For $t_1 \leq t \leq t_2$, the induced emf in the loop is

$$\varepsilon = \frac{-d\varphi}{dt} = -B\frac{dA}{dt} = -BvW$$

where A is the area of the loop in the region $y < 0$. The induced current in the loop is

$$I = -\frac{BvW}{R} \quad \text{(clockwise)}.$$

The magnetic force on the loop is

$$F = WIB = \frac{-B^2vW^2}{R} \quad \text{(upward)}.$$

The equation of motion is

$$m\dot{v} = mg - \frac{B^2 v W^2}{R}$$

from which we find

$$v = \frac{mRg}{B^2W^2} + \left(gt_1 - \frac{mRg}{B^2W^2}\right)e^{-\frac{B^2W^2}{mR}(t-t_1)}.$$

c. For $t_2 \leq t$, the only force exerting on the loop is the gravitational force. The velocity of the loop is

$$v = \frac{mRg}{B^2W^2} + \left(gt_1 - \frac{mRg}{B^2W^2}\right)e^{-\frac{B^2W^2}{mR}(t_2-t_1)} + g(t - t_2).$$

d. If the wire used in making the loop is N times larger than that used in the above case, we have the relations $R' = NR$, $W' = NW$, and $m' = Nm$. Substituting R', W', and m' into the above equation we find

$$\frac{m'R'}{(W')^2} = \frac{N^2 mR}{(NW)^2} = \frac{mR}{W^2},$$

therefore

$$t_1' = t_1 \text{ and } v_1' = v_1 \text{ for } t \leq t_1 = t_1'$$

and

$$v' = \frac{mRg}{B^2W^2} + \left(gt_1 - \frac{mRg}{B^2W^2}\right)e^{-\frac{B^2W^2}{mR}(t-t_1)} = v \text{ for } t_1 < t < t_2.$$

Therefore the motion is not affected.

(IX-6)

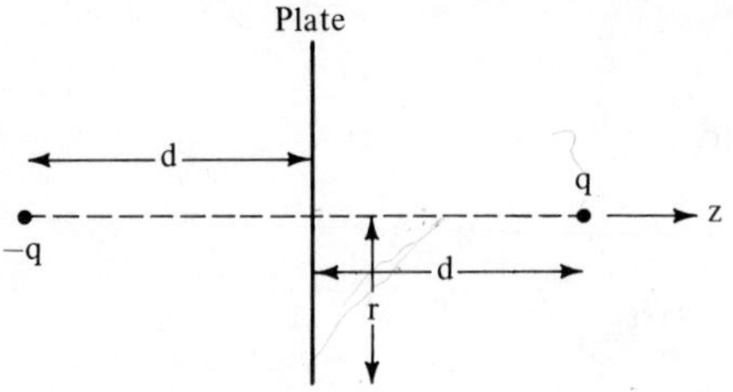

If the plane is replaced by a point charge -q located on the z-axis a distance d beyond the plane, the potential at the plane due to these two charges vanishes. The potential at infinity would also be zero since the two charges are finite. Therefore the potential due to the new system satisfies all of the boundary conditions of the old system and at the same time it is the solution of the Laplace equation. Since the solution of the Laplace equation with given boundary conditions is unique, we see that the field on the right side of the plane is the same as that produced by the two charges. We have

$$V = \frac{q}{4\pi\epsilon_o \sqrt{(z-d)^2 + r^2}} + \frac{-q}{4\pi\epsilon_o \sqrt{(z+d)^2 + r^2}} \qquad \text{for } z \geq 0.$$

The density of surface charge, derived from Gauss' law is

$$\sigma = \epsilon_o E = -\epsilon_o \frac{\partial V}{\partial z}\bigg|_{z=0}$$

or

$$\sigma = \frac{-qd}{2\pi(d^2 + r^2)^{3/2}}.$$

(IX-7) The total energy of the system is

$$W = \frac{1}{2}\epsilon_o E^2 \times (\text{vol.})$$

$$= \frac{1}{2}\epsilon_o E^2 Ax$$

where $A = 1m^2$, $x = 0.01m$, $E = 10^5$ volts/m and $\epsilon_o = 8.8 \times 10^{-12}$ farads/m. The changes in electrical potential energy when x is increased by dx is

$$dW = \frac{1}{2}\epsilon E^2 A dx.$$

Since the mechanical work done on the field must equal the increase in potential energy, we get

$$Fdx = \frac{1}{2}\epsilon E^2 A dx$$

and then

$$F = \frac{1}{2}\epsilon E^2 A$$

$$= \frac{1}{2} \times 8.8 \times 10^{-12} \times (10^5)^2 \times 1$$

$$= 4.4 \times 10^{-2} \text{ newtons.}$$

b. According to Gauss's law we have

$$\sigma = \epsilon_o E$$

$$= 8.8 \times 10^{-12} \times 10^5$$

$$= 8.8 \times 10^{-7} \text{ coul/m}^2.$$

(X-1)

a. Copper is a normal metal; the conductivity increases as temperature decreases. The cause of the increase in conductivity is that when temperature decreases, the thermal energy of the electrons and the atoms in the metal decreases. Therefore the number of collisions between electrons and between electrons and atoms decreases. Thus the current encounters less resistance and conductivity goes up.

b. Niobium is a superconducting metal; the conductivity becomes infinite as temperature goes below the transition temperature T_c. A simple explanation of superconductivity is that, below a certain temperature, the electrons in the niobium are locked together through the interaction with phonons. Since the electrons move collectively, the collisions between them are negligible. Thus the current will not be attenuated by a large number of collisions which convert the electric energy into heat in a nonsuperconductor.

c. Germanium is a semiconductor. At low temperature most electrons are in the low energy states, and the mean binding energy becomes much larger than the thermal energy (3/2kT). The number of conducting electrons decreases; therefore conductivity decreases as temperature decreases.

(X-2)

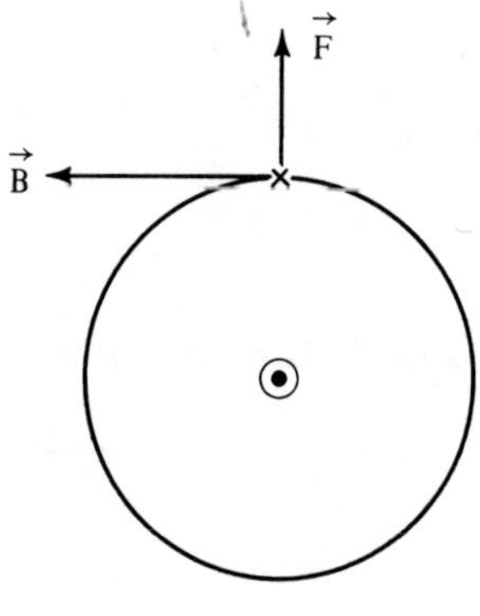

a. Calculation is similar to problem (VII-1):

$$L = \frac{\mu_0}{2\pi} \ell n \frac{20}{10} = 1.4 \times 10^{-7} \text{ henries.}$$

b. As shown in the above figure, the force on the outer conductor is in the radical direction. The force tends to burst apart the cylinder. The general rule is that when the system is connected to an energy source (e.g., a battery), the potential energy, as well as the self-inductance L (since $V = \frac{1}{2}LI^2$), tends to increase.

(X-3)

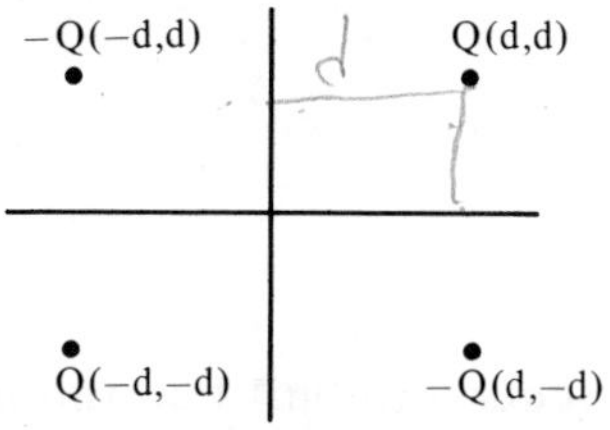

As far as the field in the first quadrant is concerned, the two conducting planes can be replaced by three point charges as shown in the figure. The electric potential in the first quadrant is,

$$V(x, y) = Q\left(\frac{1}{\sqrt{(x-d)^2 + (y-d)^2}} + \frac{1}{\sqrt{(x+d)^2 + (y+d)^2}} - \frac{1}{\sqrt{(x-d)^2 + (y+d)^2}} - \frac{1}{\sqrt{(x+d)^2 + (y-d)^2}}\right)$$

and $V(x, y) = 0$ elsewhere.

(X-4)

a. According to Ampere's law, we find the magnetic field to be

$$B = \mu_o \frac{Ni}{\ell} \tag{1}$$

where $N/\ell = 10,000$ turns per meter and $i = 100$ amp. Therefore

$$B = 4\pi \times 10^{-7} \times 10,000 \times 100 = 4\pi \times 10^{-1}$$

$$= 1.257 \text{ webers/meter}^2. \tag{2}$$

b. The total flux across any section of the ring is

$$\varphi = BA = 1.257 \times [\tfrac{\pi}{4}(10^{-2})^2] = \pi^2 \times 10^{-5}. \tag{3}$$

The self-inductance is

$$L = \frac{N\varphi}{i} = \frac{10^4(\pi^2 \times 10^{-5})}{10^2} = \pi^2 \times 10^{-3}$$

$\simeq 10^{-2}$ henries. (4)

c. The induced emf is

$$\varepsilon = -L\frac{di}{dt}. \tag{5}$$

The electrical energy needed to build up the field is,

$$E = -\int_0^t \varepsilon i dt = L\int_0^i i di = \frac{Li^2}{2} \sim 50 \text{ joules}. \tag{6}$$

d. The energy in the magnetic field is

$$E = \frac{B^2}{2\mu_0} A\ell = \frac{(4\pi \times 10^{-1})^2}{8\pi \times 10^{-7}}[\frac{\pi}{4} \times 10^{-4} \times 1] = \frac{\pi^2}{2} \times 10$$

$\simeq$ 50 joules.

(X-5) The initial charges on the two condensers C_1, C_2 are C_1V and C_2V, respectively. After they are connected, the total charges of the two condensers, Q_f, is $(C_1 - C_2)V$. The final charge on C_1 is

$$\frac{(C_1 - C_2)VC_1}{C_1 + C_2}$$

and the final charge on C_2 is

$$\frac{(C_1 - C_2)VC_2}{C_1 + C_2}.$$

The electrostatic energy before connection is

$$E_i = \frac{1}{2}(C_1 + C_2)V^2.$$

The electrostatic energy after connection is

$$E_f = \frac{1}{2}\frac{Q_f^2}{(C_1 + C_2)} = \frac{1}{2}\frac{(C_1 - C_2)^2 V^2}{(C_1 + C_2)}.$$

Therefore the energy loss is

$$\Delta E = E_i - E_f = \frac{V^2}{2(C_1 + C_2)} ((C_1 + C_2)^2 - (C_1 - C_2)^2)$$

$$= \frac{2V^2 C_1 C_2}{(C_1 + C_2)}.$$

HEAT, STATISTICAL MECHANICS, AND OPTICS

(I-1) (5 points) Estimate the specific heat of a one-cent coin.

(I-2) (5 points) Estimate the extent of polarization of atomic hydrogen in the earth's magnetic field at Berkeley for S.T.P. conditions.

(I-3) (10 points) The molecules of a gas have two states of internal energy with statistical weights g_1, g_2 and energies 0, ϵ, respectively. Calculate the contribution of these states to the specific heat of the gas.

(I-4) (15 points) A transmitting antenna for radio waves of 5-m wavelength is located on the Cliffs of Dover overlooking the English Channel, 200 m above the surface of the Channel. An airplane flying just above the surface of the water 20 km away cannot receive signals from the antenna and thus does not reflect echo signals back to the transmitting site. Why is this so? At certain altitudes, on the other hand, the plane will reflect exceptionally strong echoes. What are these altitudes? (This physical effect was instrumental in the winning of the Battle of Britain).

(I-5) (20 points) It is desired to take a picture of a distant yellow object with a pinhole camera in which the distance from pinhole to film is D. Approximately what should be the diameter of the pinhole if the picture is to be of maximum sharpness? In cameras of this optimum design and using a film of fixed speed, how will the exposure time for the picture depend upon D?

(I-6) (20 points) A one-liter bulb at room temperature contains hydrogen gas at a pressure of 10^{-4} Torr (i.e., 10^{-4} mm of Hg). At a given instant, t = 0, a filament of area 0.2 cm^2 is suddenly heated to incandescence. Under these conditions, hydrogen molecules striking the filament are dissociated. Neutral H atoms so produced stick to the walls of the bulb when they strike there. Approximately how long is the mean free path for hydrogen molecules at the starting pressure? Derive an expression for the pressure as a function of time, t. How long does it take for the pressure to drop to a value 10^{-7} Torr. Neglect any changes in gas temperature induced by turning on the filament.

(I-7) (25 points) The following data apply to the triple point of water:

Temperature: 0.01° C
Pressure: 4.58 mm of Hg
Specific volume of solid: 1.0907 cm^3/g
Specific volume of liquid: 1.0001 cm^3/g
Heat of fusion: 80 calories/g
Heat of vaporization: 596 calories/g

Sketch a P-T diagram for water which need not be to scale, but which should be qualitatively correct. Then consider what happens when the pressure is reduced slowly from some high value upon an amount of pure water enclosed in a cylinder and maintained at the temperature -1° C. Two phase changes will occur. Describe what these phase changes are and calculate the pressures at which they occur.

(II-1) (10 points) Numerical answers should be correct within one order of magnitude.

(a) What is the number of molecules/cm^3 of air at S.T.P.?

(b) What is the mean free path of the N_2 molecule in air at S.T.P.?

(c) At what temperature is the root-mean-square velocity of the N_2 molecule equal to the escape velocity from the surface of the earth?

(II-2) (10 points)

(a) What is the velocity of sound in air at S.T.P.?

(b) What is the efficiency of the most efficient cyclic heat engine operating between heat reservoirs at temperature T_1 and T_2 where $T_1 > T_2$?

(c) Adiabatic demagnetization of a paramagnetic salt usually results in what physical phenomena?

(II-3) (10 points)

(a) Describe the characteristics of a first-order phase transition in a substance.

(b) What is the Boltzmann relation connecting the entropy of a system in a given state and the probability of occurrence of the state?

(c) The specific heat of copper is approximately:

(i) 10^{-2} cal/gm° C

(ii) 10^{-1} cal/gm° C

(iii) 1 cal/gm° C

(iv) 10 cal/gm° C

(II-4) (10 points) The equilibrium temperature of a 100-gram object is observed as a function of input power (see table below). Later the decrease in temperature is observed with no power input (see table on right). Assuming that all parts of the object are at the same temperature during the cooling, deduce the heat capacity of the object in joules/gram° C.

Equilibrium temperature		Temperature vs time at zero power input	
Temp.	Power	Time	Temp.
20° C	0 watts	0 sec	40.00° C
25°	1	50	35.74
30°	2	100	32.39
35°	3	150	29.76
40°	4	200	27.68
		250	26.05
		300	24.74
		350	23.75
		400	22.95
		450	22.32
		500	21.83

(II-5) (5 points) N particles are distributed among three states having energies $E = 0$, $E = kT$, and $E = 2kT$. If the total equilibrium energy of the system is 1000 kT, what is the value of N?

(II-6) (10 points) A piece of ice at 0° C and a beaker of water at 0° C are placed side by side in a small bell jar from which all air has been removed. If the ice, water, and vessel are all individually maintained at 0° C by accurate thermostats, describe the final equilibrium state in the apparatus if the system is kept at 0° C. Justify your answer. (The triple point of water is 0.0098° C and 4.579 mm of Hg. The critical point is 374° C and 218 atmospheres.) What is the final state if the temperature is not kept at 0° C and the system is thermally isolated?

(II-7) (3 points) Estimate the minimum lens diameter required to resolve objects one foot apart at a distance of 100,000 ft.

(II-8) (3 points) A point source of light is viewed through a plate of glass. Does the source appear closer, farther away, or at the same distance?

(II-9) (3 points) A projector makes an image of a slide on a screen 15 feet from the lens. If the 1-inch dimension of the slide is magnified to 2 feet, what is the lens focal length?

(II-10) (3 points) Where is the image located that is viewed in looking through a microscope?

(II-11) (3 points) A monochromatic beam of light passes through a narrow slit and the Fraunhofer diffraction pattern is observed. By what factors do the intensity of the center of the pattern and the total energy transmitted change when the slit width is doubled?

(II-12) (3 points) Consider Newton's rings viewed in transmitted and reflected light from a convex lens in contact with a flat glass plate. Is the intensity of the reflected fringe system more intense, less intense, or equally intense as the transmitted fringe system?

(II-13) (3 points) Circularly polarized light is passed through a quarter-wave plate. What is the general polarization state of the outgoing light?

(II-14) (3 points) A circular opening of variable radius R is placed a distance D from a screen. As the opening is enlarged from zero radius, for what radius does the intensity of light of wavelength λ go to zero for the first time?

(II-15) (3 points) A source and screen are fixed in place a distance ℓ apart. A thin lens is placed between them at a position such that the source is focused on the screen. For what ranges of lens focal lengths are there two, one, and no such position?

(II-16) (3 points) Give quantitative estimates for the magnitude and slope of the index of refraction curve for glass in the visible region.

(II-17) (3 points) In question (II-16) a particular lens makes a 1/2-inch image of a 1-inch source. The lens is now moved to the other position where an image of the source is in focus on the screen. What is the image size?

(II-18) (3 points) A star is viewed by eye at night. How large is the image formed on the retina?

(II-19) (3 points) For what position does a plane mirror produce a real image?

(II-20) (3 points) What is the Rayleigh Criterion?

(II-21) (3 points) Two thin lenses of focal length f are placed a distance ℓ apart. Are there any values of ℓ? If so, what are the values of ℓ that lead to a negative focal length for the combination?

(III-1) (10 points) In the system shown in the diagram, rays from a point object are first deviated by a prism and then focused by a thin lens of focal length f. The prism is made of glass of index n and has a small angle α (i.e., small angle approximations are valid).

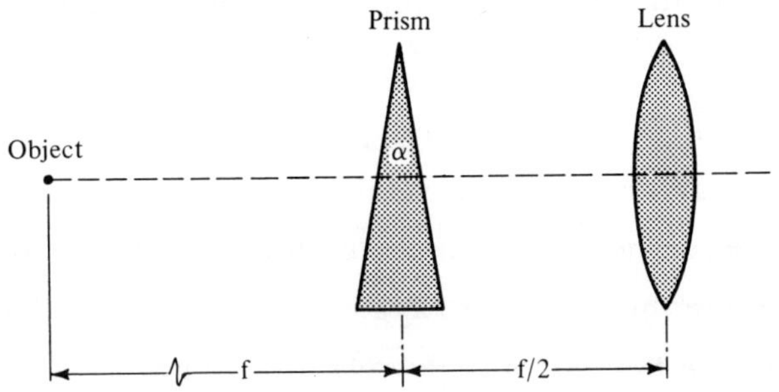

(a) Calculate the deviation angle of rays hitting the prism at nearly normal incidence (i.e., perpendicularly to one of the faces).

(b) If the distances between object, prism, and lens are as shown in the diagram, locate the image position both along and transverse to the axis.

(III-2) (10 points) A stack of N glass plates of refractive index n and thickness t are assembled with each plate projecting a distance s beyond the one following it:

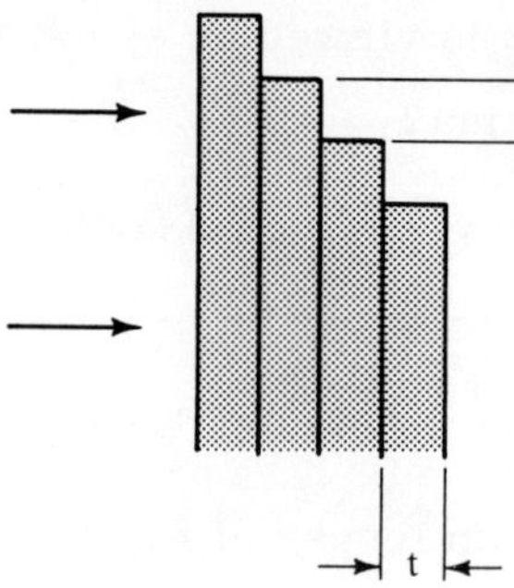

The stack is illuminated from the left by light of wavelength λ. Considering light coming from each step in the assembly, what is the condition for constructive interference beyond the stack for small angles with respect to the incident beam? What is the order of interference for $n = 1.5$, $t = 0.5$ cm and $\lambda = 5000$ Å? Calculate the angular dispersion and resolving power for a stack of 40 plates of this index and thickness, assuming that the refractive index varies only slightly.

(III-3) (10 points) Right circularly polarized light is passed through two identical quarter-wave plates whose optic axes make an angle θ with each other. The refractive index for the extraordinary rays is less than that for the ordinary rays.

(a) What is the outgoing polarization state if $\theta = 0°$?

(b) for $\theta = 45°$?

(c) for $\theta = 90°$?

(d) Would you say that the two quarter-wave plates are identical in their effect to a single half-wave plate?

If any of the above answers is circular polarization, state whether right or left circular polarization. If any answers involve plane polarization, state the direction of polarization. (Note: In right circular polarization, the electric vector rotates clockwise as seen by an observer looking toward the source.)

(III-4) (20 points) Suppose you are given the following unlabelled optical devices:

(a) Two linear polarizers

(b) A quarter-wave plate

(c) A half-wave plate

(d) A circular polarizer

Describe in detail how you would identify each of the elements without the aid of any other optical instruments (except a lamp and a screen). What if in (a) you had only one linear polarizer?

(III-5) (20 points) Consider a sample of N magnetic atoms, each with spin 1/2. The system is known to be ferromagnetic at very low temperatures; thus as $T \to 0$ all the spins are aligned. At sufficiently high temperatures, the spins are randomly oriented. Neglect all other degrees of freedom but the spin orientation.

(a) Define the entropy of the system:

(i) In statistical terms

(ii) In terms of the specific heat and the (spin) temperature.

(b) Show that the specific heat C(T) must satisfy the equation

$$\int_0^{\infty} \frac{C(T)dT}{T} = kN \ln 2$$

irrespective of the details of the interactions bringing about ferromagnetic behavior, and irrespective of the detailed dependence of C on T.

(III-6) (20 points) A thin-walled vessel of volume V is kept at constant temperature T. A gas slowly leaks out of the vessel through a hole of area A into surrounding vacuum. Find the time required for the pressure in the vessel to drop to 1/e of its original value.

(III-7) (10 points) The surface temperature of the sun is $T_0 = 5500°K$, its radius is $R = 7 \times 10^{10}$ cm, the radius of the earth is $r = 6.4 \times 10^8$ cm, and the distance between sun and earth is $D = 1.5 \times 10^{13}$ cm. Assume that earth and sun both absorb all EM radiation incident on them, and that the earth is in a steady state with T constant. Calculate T from the parameters given.

(IV-1) (10 points)

(a) A liter of N_2 gas at atmospheric pressure and 0° C in a rigid cylinder is raised to 100° C by placing it in contact with an infinite reservoir at 100° C. What are the changes in entropy of the N_2 and the universe?

(b) Assuming that one wall of this cylinder is allowed to act like a piston, describe a means of raising the gas temperature to 100° C (with the final volume of 1 liter) such that the entropy change $\Delta S = 0$ for the universe. (Gas constant $R \simeq$ cal/mole.)

(IV-2) (10 points) Show how a third lens may be inserted in the optical instrument shown to increase the light-collecting efficiency without changing the image and object positions.

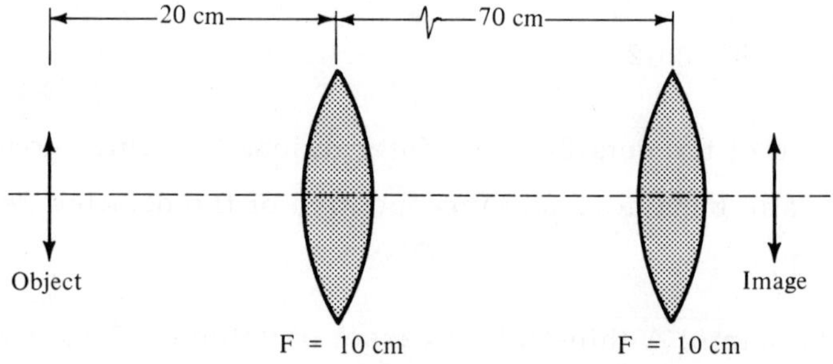

Indicate the position and focal length of the additional lens.

(IV-3) (5 points) Newton's rings are formed by reflection in the air film between a plane surface and a spherical surface of radius 50 cm. If the radius of the third bright ring is 0.09 cm and of the twenty-third 0.25 cm, what is the wavelength of the light used?

λ = __________ cm.

(IV-4) (15 points) Given a box of air with wall temperature T_1. There is a small amount of water lying on the floor of the box, at equilibrium. The volume is V_1 and the pressure is p_1. The volume of the box is now slowly increased, keeping the temperature T_1 constant. When the volume has doubled to $V_2 = 2V_1$ the water on the floor has essentially disappeared. If $p_1 = 3$ atmospheres and $p_2 = 2$ atmospheres, what is the wall temperature T_1? If the volume is doubled once more to $V_3 = 2V_2$, what will be the pressure p_3? If $V_2 = 44.8$ liters, what is the mass of the water (liquid or vapor) in the box? What is the mass of air?

(IV-5) (10 points)

(a) Calculate the change in entropy when one mole of an ideal gas is allowed to expand freely into double its original volume.

(b) What is the entropy change when one mole each of two distinct noninteracting ideal gases are allowed to mix, starting with equal volumes and temperatures?

(c) What entropy change is there when the valve connecting two equal-volume and -temperature bulbs of the same gas is opened?

(IV-6) (20 points) Suppose right-handed circularly polarized light (defined to be clockwise as the observer looking toward the on-coming wave) is incident on an absorbing slab. The slab is suspended by a vertical thread. The light is directed upwards and hits the underside of the slab.

(a) If the circularly polarized light beam has 1 watt of visible light of average wavelength 6200 Å, and if all of this light is absorbed by the slab, what is the torque, τ, exerted on the slab? (Give the answer to (a) in dyne-cm and the answer to the remaining parts in units of τ.)

(b) Suppose that instead of an absorbing slab you use an ordinary silvered mirror surface, so that the light is reflected back at 180° to its original direction. What is the torque now?

(c) Suppose that the slab is a transparent half-wave plate. The light goes through the plate and doesn't hit anything else. What is the torque? (Neglect reflections at the surfaces of the slab.)

(d) Suppose the slab is a transparent half-wave plate with the top surface silvered, so the light goes through the half-wave plate, reflects from the mirror, and returns through the plate. What is the torque?

(e) The slab is a transparent half-wave plate. Above the slab is a fixed quarter-wave plate (i.e., not attached to the slab) silvered on the top surface so as to reflect light back through the slab. What is the torque exerted on the slab?

(IV-7) (20 points) Give short answers:

(a) A community decides to limit its population by requiring that couples stop producing children after their first boy is born. If 51% of the children born are boys under unlimited conditions, what fraction of the children born under the limited scheme will be boys? Why?

(b) Given an apparatus for emitting and detecting electromagnetic waves at all frequencies, design an experiment to determine whether an object under consideration is a black body.

(IV-8) (10 points) A sample of protons (in some crystal) is in a uniform external magnetic field B. When the sample is irradiated by electromagnetic waves of suitable polarization, the maximum rate of power absorption (due to flipping of proton spins) occurs at a frequency of 100 Mc. What is the fractional polarization P of this sample of proton spins, with this field B, and assuming room temperature? Note: P is defined as N(up) - N(down) divided by N(up) + N(down).

(V-1) (5 points) Sketch in Fig. 1 (roughly to scale) the specific heat for hydrogen gas.

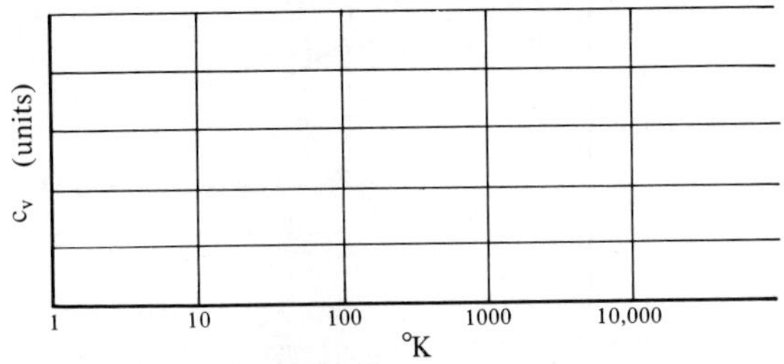

(V-2) (5 points) A 100-ohm resistor carrying a current of I amps is maintained at a constant temperature of 30° C by a heat bath. What is the rate of entropy increase of the resistor?

__________ joules/°K/sec

(V-3) (5 points) At low temperatures, He^4 is used as a heat-exchange medium. At similar temperature and pressure, would the heat exchange using He^3 be smaller, the same, or larger?

____

(V-4) (5 points) A gas-filled tube is whirled about one end with angular velocity ω. What is the equilibrium density distribution in the tube?

$\rho(x) =$ __________

(V-5) (5 points) A sphere (moment of inertia I) is suspended from a thin filament (restoring force C) in a gas at temperature T. What are the mean values for:

$\overline{\theta^2}$ __________ $\overline{\dot{\theta}^2}$ __________

If the pressure of the gas is now reduced to $10^{-6}\ p_o$:

$\overline{\theta^2}$ __________ $\overline{\dot{\theta}^2}$ __________

(V-6) (5 points) What is the efficiency for a reversible engine operating around the cycle illustrated?

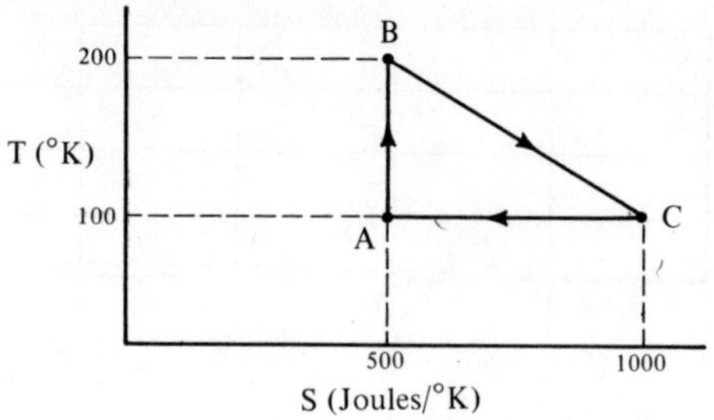

η = ________ %

(V-7) (5 points) It is desired to cool a gas by expansion through a porous plug. Should the initial conditions be chosen so that $\partial h/\partial p)_T$ is greater than, equal to, or less than zero?

(V-8) (5 points) A particle undergoes Brownian motion in a gas at temperature T and pressure p. The mean-square displacement is inversely proportional to the viscosity. Will the mean-square displacement be smaller, larger, or remain the same if the pressure is halved?

$\overline{d^2}_{(p=1/2p_0)}$ = ________ $\overline{d^2}_{(p_0)}$

(V-9) (5 points) A radiation gas of temperature T fills a cavity of volume V. The system expands adiabatically and reversibly to a volume equal to 8V. By what factor does the temperature change?

T_f = ________ T_1

(V-10) (5 points) A thermally insulated system consists of a mass suspended from a spring, initially displaced a distance A from its equilibrium position. The mass is released and gradually comes to rest due to internal damping. Does the entropy of the universe change? If so, by how much?

ΔS = ________

(V-11) (5 points) Assume that the electrons inside a 1-cm cube of copper at zero °K behave like a completely degenerate Fermi gas. Indicate the characteristic features of the energy distribution expected for the electrons.

$dN/dE \propto$ __________

$E_{max} \propto$ __________

(V-12) (10 points) Compute the minimum amount of work (joules) required to freeze one liter of water originally at temperature $T = 20° C$. Consider the heat reservoir to be at 20° C. The heat of fusion of water is 80 cal/gm.

(V-13) (15 points) Calculate the boiling point of water at an altitude of 1,000 feet. State clearly the steps in the calculation and the approximations made. The heat of evaporation of water at N.T.P. is 540 cal/gm.

(V-14) (10 points) The specific heat capacities of liquid water and ice at atmospheric pressure and for several degrees Celsius below the ice point are given by the equations

$$c_p(\text{water}) = 4222 - 22.6t$$

$$c_p(\text{ice}) = 2112 + 7.5t$$

in joules per kilogram-degree. What is the specific entropy in joules/degree for each system at -10° C? (Choose a convenient reference.) Which of these systems may be considered in the more ordered state?

(V-15) (10 points) Two lenses, with focal lengths 20 cm and 30 cm, respectively, are separated by a distance of 10 cm. An object of 5-cm length is put in front of the first lens perpendicular to the optical axis. The distance from the object to the first lens is 30 cm. Find the size of the image behind the lens system. State also whether the image is erect or inverted.

(VI-1) (10 points) Interference fringes are produced by a thin, wedge-shaped film of plastic of refractive index 1.4. If the angle of the wedge is 20 seconds of arc and the distance between fringes is 0.25 cm, find the wavelength of the monochromatic light incident perpendicularly on the plastic.

(VI-2) (10 points) Two identical bodies of constant heat capacity C_p are used as reservoirs for a heat engine. Their initial temperatures are T_1 and T_2, respectively. Assuming that the bodies remain at constant pressure and undergo no change of phase, derive an expression for the maximum work obtainable from the system.

(VI-3) (10 points) A cubical container, 1 cm on a side, contains helium at standard temperature and pressure. Estimate, to within a factor of ten,

(a) the number of collisions per second experienced by any one helium atom

(b) the number of collisions that the atoms make with one wall of the container in one second.

(VI-4) (10 points) In a perfect gas of electrons, the mean number of particles occupying a single-particle quantum state of energy E_i is:

$$\overline{N}_i = \frac{1}{e^{\beta(E_i-\mu)} + 1} \qquad \left(\beta \equiv \frac{1}{kT}\right)$$

(a) Obtain a formula which could be used to determine μ in terms of the particle density n, the temperature T, and various constants.

(b) Show that the expression above reduced to the Maxwell-Boltzmann distribution in the limit $n\lambda^3 \ll 1$, where λ is the thermal de Broglie wavelength.

(c) Estimate the quantities in the inequality for the electron gas in a metal at room temperature, and show that Maxwell-Boltzmann statistics may not be used for that gas.

(VI-5) (10 points) The index of refraction $n(\omega)$ of an ionized medium, for an electromagnetic wave (of angular frequency ω) propagating parallel to a constant magnetic field, is given by

$$n_{\pm}^{2}(\omega) = 1 - \frac{K}{\omega(\omega \pm \Omega)} ;$$

where K and Ω are constants, and $\pm$ refer to circular polarization in the opposite and the same sense, respectively, as the gyration of electrons in the magnetic field. Consider a slab of this medium, of thickness L, with the magnetic field normal to the slab. A plane-polarized monochromatic electromagnetic wave is normally incident on the slab. Neglecting reflection from the slab surfaces, find the amount of Faraday rotation, i.e., the angle through which the plane of polarization is rotated in traversing the slab, in terms of the given constants. Assume ω sufficiently large that $n_{\pm}^{2} > 0$.

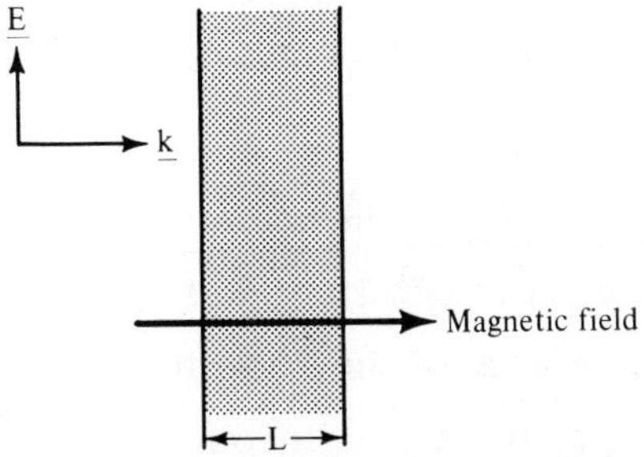

(VI-6) (5 points) While an aquarium is being filled with water, a motionless fish looks up vertically through the surface of the water at a monochromatic plane-wave source of frequency ν. If the index of refraction of the water is n and the water level rises at a rate dh/dt, calculate the shift in frequency which the fish observes.

(VI-7) (5 points) A quartz crystal has indices of refraction $n_e = 1.55379$ and $n_o = 1.54225$ for light of wavelength 5829.90 Å. How thick must a parallel plate of quartz be cut, with the optic axis parallel to the surfaces, to make a quarter-wave plate?

(VI-8) (10 points) The resolving power of a Fabry-Perot interferometer is given by

$$\frac{\lambda}{\Delta\lambda} = \frac{m\pi r}{1 - r^2}$$

where m is the order of interference and r^2 is the reflectance of the mirrors (and r^2 is not much less than unity). A gas laser has mirrors at its ends and produces a spectrum of lines which are separated according to various orders of interference between the two mirrors. Assume that the laser is 30 cm long, and the wavelength is 6000 Å.

(a) Find the separation between two neighboring lines given out by the laser around 6000 Å.

(b) Calculate the reflectance of a Fabry-Perot interferometer with 1.5-cm plate spacing such that this F.P. can just resolve the line spacing found in part (a).

(VI-9) (20 points) Given the 3-media shown, an incident plane electromagnetic wave of intensity I_o enters normal to the interface from medium 1.

(a) Find the intensity of the wave transmitted into medium 3. (Medium 3 extends indefinitely downward.)

(b) Find the intensity of the wave in medium 1 which returns in the direction opposite to the incoming wave.

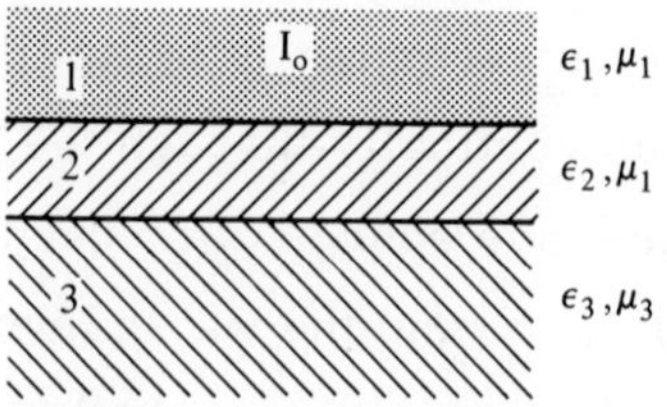

(VI-10) (5 points) Considering the limitations imposed by diffraction, calculate the greatest distance ℓ at which the human eye can distinguish the two headlights of an auto.

ℓ = __________ meters

(VI-11) (5 points)

(a) For maximum polarization of light by reflection from a plane dielectric, what is the angle between reflected and refracted rays?

(b) What is the angle between refracted ray and the normal to the dielectric surface?

(VII-1) (3 points) A double slit is illuminated first by red light and then violet light. Which color gives the wider interference pattern beyond the double slit?

(VII-2) (3 points) If an object is placed at the center of curvature of a concave spherical mirror, where is the image located?

(VII-3) (3 points) In what plane is the E-vector of light polarized after being reflected from glass at Brewster's angle?

(VII-4) (3 points) A beam of light, initially converging toward a point P, is made to pass through a plate of glass before reaching P. If the plate of glass is erected perpendicular to the axis of the beam of light, in which direction is the point of convergence shifted?

(VII-5) (3 points) An object is placed on the axis inside the focal point of a thin, positive lens. Where is the image located?

(VII-6) (3 points) Within a given order of a diffraction grating, which color (red or violet) is located closest to the next higher order?

(VII-7) (3 points) For a thin positive lens, is the focal length for red light longer or shorter than that for violet light?

(VII-8) (3 points) An object is placed on the axis outside the focal point of a thin, negative lens. Is the image real or virtual?

(VII-9) (3 points) What type of aberration is present in lenses and absent from mirrors?

(VII-10) (3 points) What types of phenomena provide direct

experimental proof of the transverse nature of light?

(VII-11) (3 points) For maximal polarization by reflection, what is the angle between the reflected and refracted rays?

(VII-12) (3 points) For a given thick lens, what condition must a ray not parallel to the axis satisfy to go through the lens undeviated?

(VII-13) (3 points) What is the range of frequencies for visible light?

(VII-14) (3 points) A monochromatic beam of light goes from vacuum into a medium of refractive index n. What is the relation between the frequencies of the incident and refracted waves? What is the relation between their wavelengths P?

(VII-15) (13 points) A paperweight in the form of a glass hemisphere is placed, flat surface down, on a page of a book. Calculate the position and magnification of the image of the type at the center of the flat surface of the hemisphere. Illustrate this result with a ray diagram, using an arrow to represent a letter of type on the page of the book. Take the index of refraction of the glass as 1.50.

(VII-16) (15 points) The diagram shows a double slit experiment in which monochromatic light of wavelength λ from a distant source is incident upon two slits, each of width w ($w \ll \lambda$), and the interference pattern is viewed on a distant screen.

A thin piece of glass of thickness δ, index of refraction n, is placed between one of the slits and the screen, and the intensity at the central point C is measured as a function of thickness δ. If the intensity for $\delta = 0$ is given by I_o:

(a) What is the intensity at point C as a function of thickness δ?

(b) For what values of δ is the intensity at C a minimum?

(c) Suppose that the width of one of the slits is now increased to $2w$, the other width remaining unchanged. What is the

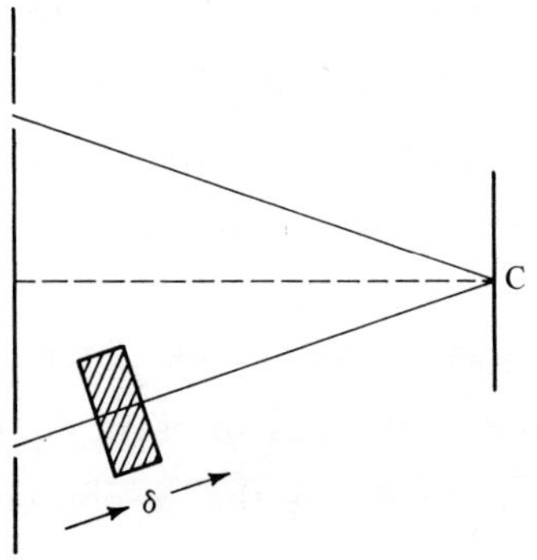

intensity at point C as a function of δ ?

Assume that the glass does not absorb any light.

(VII-17) (20 points) Sound velocity is about 330 m/sec. Consider a whistle that emits a steady note at 3300 cps. Far from the whistle is a circular slab of perfect sound-absorbing material, of diameter 2 m. The slab is oriented perpendicular to the direction from whistle to center of the slab.

If your head is behind the center of the slab, you cannot hear the whistle. If you go very far downstream from the slab and locate your head so that the slab "hides" the whistle and is centered on line of sight from you to whistle, you might think you could not hear the whistle. Instead, if you are far enough downstream, you can hear the whistle as plainly with the slab in place as with the slab removed.

(a) Why is that? (Note: Neglect effect of ground, of buildings, trees, etc. Let the whistle be in free space--but in air. Both you and the slab are suspended in free air by balloons or something.)

(b) In as simple a way as possible, derive a formula that you then use to answer the following question: about how far downstream (in meters) from the slab must you be for the insertion of the slab to reduce the sound intensity by about a factor of 2? We want to know this distance only to within a factor of 2.

(VII-18) (10 points) A laser beam (λ = 6000 Å) on earth is focused, by a telescope of lens (or mirror) diameter 2 m, on a crater on the moon (distance 400, 000 km). How big is the spot on the moon? Neglect the effect of the earth's atmosphere.

(VIII-1) (5 points) A gas of molecules of mass m is enclosed in a stationary tank. Suppose the gas to be in equilibrium at the absolute temperature T. If V_x denotes the x-component of the velocity of a molecule, find the following mean values:

V_x = __________, $\overline{V_x^2}$ = __________, $\overline{V_x^3}$ = __________

(VIII-2) (5 points) In the Millikan oil drop experiment, the terminal velocity with which the oil drop falls is inversely proportional to the viscosity of the air. If the temperature of the air is raised, does the terminal velocity of the drop increase, decrease, or remain the same?

(VIII-3) (5 points) Fifty grams of milk at temperature T_1 are slowly poured into 250 grams of coffee at temperature T_2. Assume the specific heats of both liquids to be the same as that of water. What is the final temperature T_f after equilibrium is reached? What is the total entropy change ΔS?

(VIII-4) (5 points) An ideal monatomic gas is compressed (no heat being added or removed in the process) so that its volume is halved. What is the ratio of the new pressure to the original pressure?

(VIII-5) (5 points) The molar specific heat C of electrons in a metal is given by $C = \gamma T$, where T is the absolute temperature and γ is a constant. What is the contribution of these electrons to the molar entropy of the metal?

S = __________

(VIII-6) (5 points) The motion of atoms in thermal equilibrium at absolute temperature T produces a Doppler shift in the frequency of an emitted spectral line. If v_z is the z-component of the molecular velocity and ν_o is the intrinsic frequency of the line, then the observed frequency is given by $\nu = \nu_o(1 + v_z/c)$. What is the observed mean-square line width?

$\overline{(\nu - \nu_o)^2}$ = ________

(VIII-7) (5 points) A solid of density ρ_1 melts at pressure p and absolute temperature T to form a liquid of density ρ_2. The latent heat of melting per gram of solid is L. Find the change of entropy ΔS and the change of internal energy ΔU resulting from the melting of a gram of the solid.

ΔS = ________, ΔU = ________

(VIII-8) (5 points) An atom can be in each of two quantum states separated by an amount of energy E. Make a qualitative sketch of the specific heat C of a collection of such atoms as a function of absolute temperature T. Be sure to indicate correctly what happens in the limits of $T \to 0$ and $T \to \infty$.

(VIII-9) (5 points) Consider an ideal gas of diatomic rigid (i.e., nonvibrating) molecules. What is the molar specific heat of this gas, taking into account quantum mechanics (a) in the limit of "low" temperatures and (b) in the limit of "high" temperatures. (You must interpret for yourself the meaning of "low" and "high.")

(VIII-10) (5 points) Let $E(\lambda)d\lambda$ be the radiation emitted per second by a black body in the wavelength range between λ and $\lambda + d\lambda$. The graph shows a plot of $E(\lambda)$ vs λ for a temperature of 200° K. On the same graph, draw a plot of $E(\lambda)$ vs λ for T = 1000° K in reasonable proportion. Let λ^* denote the wavelength.

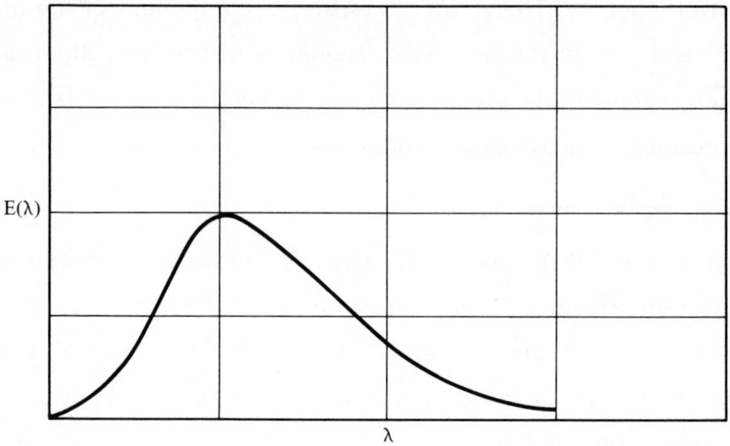

(VIII-11) (20 points)

(a) Calculate the maximum work in joules obtainable from a heat reservoir consisting of 200 kg of iron heated initially to a temperature of 1500° C, using the ocean, at 12° C, as the second heat reservoir. Assume that the specific heat capacity of the iron is constant and equal to 0.60 joules/gram-deg.

(b) Calculate the entropy change of the universe in this process.

(VIII-12) (20 points)

(a) A half-silvered mirror will reflect half the incident light and transmit the remainder. Suppose the light is plane-polarized so that the incident electric vector lies in the plane of the mirror. Calculate the phase changes appropriate to the reflected and transmitted components. (Assume mirrors of zero thickness.)

(b) A monochromatic coherent light beam, polarized as above, originates at A and passes through a rectangular array of mirrors. Mirrors C and D are totally reflecting, while B and E are half-reflecting. The optical lengths are chosen so that total destructive interference occurs at G. Conservation of energy requires that there be total constructive interference at F. Show that this actually occurs.

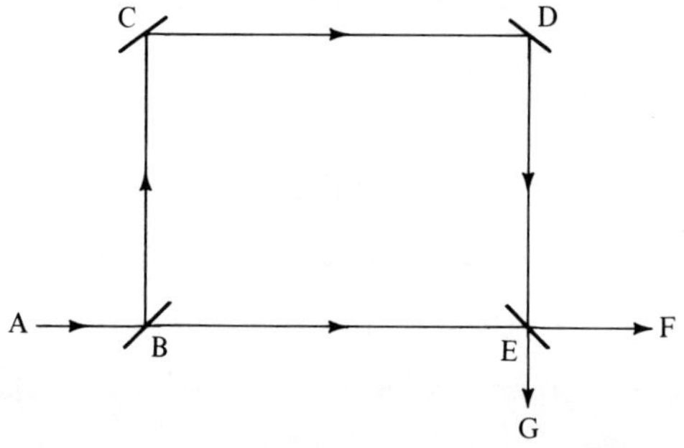

(VIII-13) (10 points) Rays from an object immersed in water ($n = 4/3$) traverse a spherical air bubble of radius R.

(a) Locate the image of an object which is very distant from the bubble (i.e., $x = \infty$ on the diagram). State whether the image is inverted or erect, real or virtual.

(b) Repeat part (a) for an object located at the left bubble surface (i.e., $x = 0$ on the diagram). Consider only rays close to a diameter of the bubble.

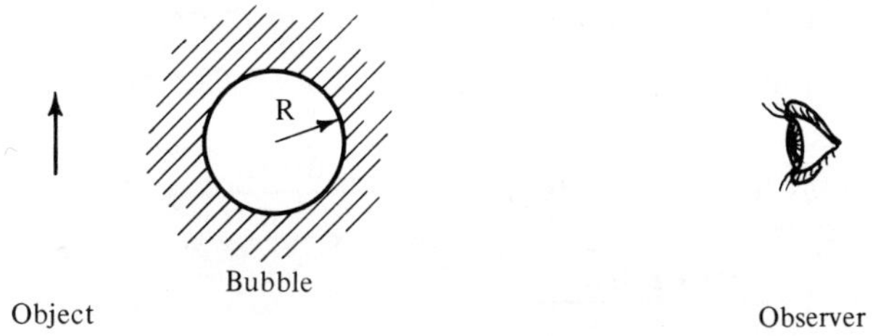

SOLUTION SET OF

HEAT, STATISTICAL MECHANICS, AND OPTICS

(I-1) The main point of physics here is the fact that at ordinary temperatures copper obeys the law of Dulong and Petit, i.e., the molar heat capacity is $3R \cong 6$ cal/mole-deg.

The atoms of copper vibrate in three dimensions about equilibrium sites. Since the potential energy of the atoms is quadratic in the x, y, z displacements from the equilibrium positions, and since the kinetic energy of the atoms is quadratic in dx/dt, dy/dt, and dz/dt, there are six degrees of freedom. The frequencies of vibration are such that $kT \gg \hbar\omega$, so that the classical law of equipartition of energy holds for each of these degrees of freedom, and the molar heat capacity is $(6)(\frac{1}{2}R) = 3R$.

Specific heat is strictly heat capacity relative to water or heat capacity per gram, since the heat capacity of water is by definition one calorie per gram degree. Thus the mass of the one-cent coin is irrelevant, but the atomic weight of copper is relevant.

$$c_V \sim 3R(\text{calories per mole degree}) \times \left(\frac{1 \text{ mole}}{63.5 \text{ gm}}\right)$$

$$= \frac{6}{63.5} \cong 0.1 \frac{\text{calories}}{\text{gram-degree}} .$$

(I-2) This is an order-of-magnitude question. The extent of polarization will be of the order of $\mu H/kT$, where μ is the Bohr magneton. A simple Boltzmann factor argument shows that the percent polarization is, in fact, $100\ \mu H/kT$.

Most physicists remember that the Bohr magneton $\sim 10^{-20}$ emu, which is easily computed from $\mu = e\hbar/2mc$. The earth's magnetic field is about 0.5 gauss. Thus at S.T.P., $\mu H/kT \sim (10^{-20})(0.5)/(1.4 \times 10^{-16})(273) \sim 10^{-7}$ so that the polarization is very small indeed.

(I-3) The formal means of solving this problem is to form the molecular partition function

$$Z = g_1 + g_2 \exp(-\epsilon/kT)$$

and to use the formula which relates the internal energy to the partition function. However we will give a solution which requires only the use of the Boltzmann factor. Let

n_1 = number of molecules in state 1

n_2 = number of molecules in state 2

then

$$\frac{n_2}{n_1} = \left(\frac{g_2}{g_1}\right)\exp(-\epsilon/kT).$$

If we consider one mode of gas then $n_1 + n_2 = N$ = Avogadro's number. Solving

$$\left(\frac{N - n_1}{n_1}\right) = \left(\frac{g_2}{g_1}\right)\exp(-\epsilon/kT)$$

or

$$n_1 = \frac{N}{1 + \frac{g_2}{g_1} e^{-\epsilon/kT}}$$

$$n_2 = \frac{N\frac{g_2}{g_1} e^{-\epsilon/kT}}{1 + \frac{g_2}{g_1} e^{-\epsilon/kT}}$$

$$U_{int} = n_1\epsilon_1 + n_2\epsilon_2 = \frac{N\epsilon \frac{g_2}{g_1} e^{-\epsilon/kT}}{1 + \frac{g_2}{g_1} e^{-\epsilon/kT}},$$

then the molar specific heat at constant volume is

$$(C_v)_{int} = \left(\frac{\partial U_{int}}{\partial T}\right)_v$$

$$= \frac{(1 + \frac{g_2}{g_1}e^{-\epsilon/kT})N\epsilon\frac{g_2}{g_1}e^{-\epsilon/kT}(\epsilon/kT^2) - N\epsilon\frac{g_2}{g_1}e^{-\epsilon/kT}\cdot\frac{g_2}{g_1}e^{-\epsilon/kT}(\frac{\epsilon}{kT^2})}{(1 + \frac{g_2}{g_1}e^{-\epsilon/kT})^2}$$

$$= \frac{\frac{N\epsilon^2}{kT^2}\frac{g_2}{g_1}e^{-\epsilon/kT}}{(1 + \frac{g_2}{g_1}e^{-\epsilon/kT})^2} = \frac{\frac{N\epsilon^2}{kT^2}g_1 g_2 e^{+\epsilon/kT}}{(g_1 e^{\epsilon/kT} + g_2)^2}$$

(I-4) This is essentially Lloyd's mirror.

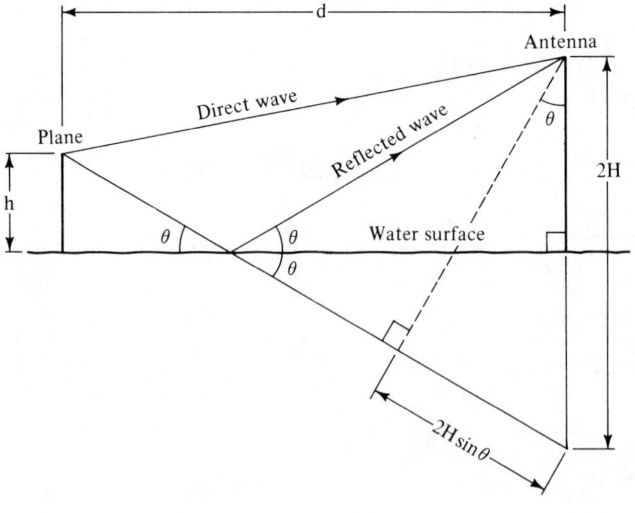

When the plane is very near the water, it receives a direct wave plus a reflected wave of the same path length. (In the above figure, the plane is shown a little bit above the water to prevent the two paths from being completely superposed.) Since there is a phase change upon reflection, at grazing angle the direct and reflected waves cancel and no signal is received.

For a sufficient height above the water, a path difference of $\lambda/2$ between the two waves will occur and we get a maximum signal. Similarly, when the path difference is $(n + 1/2)\lambda$ $(n = 0, 1, 2, \text{etc.})$ there will be maxima. For a path difference of $(n + 1/2)\lambda$ we find $(n + 1/2)\lambda \cong 2H \sin\theta$ or $H + h \cong d \sin\theta = (n + 1/2)d\lambda/2H$. Here $\lambda d/2H = (5)(2 \times 10^4)/2(200) = 250$ m. Thus maxima occur at heights of $h = ((3/2, 5/2, \text{etc.}) \times 250 - 200)$m.

(I-5) When the pinhole is quite large, a point on the object is imaged as a geometrical shadow of the front of the camera or as a disc of diameter d, the diameter of the pinhole. When the pinhole is quite small, the image of a point on the object is a diffraction disc of diameter $\simeq 2 \times (1.22\frac{\lambda}{d}) \sim 2.4\,\lambda/d$ or linear diameter $\simeq 2.4\lambda D/d$. Since we want to minimize the size of this image disc for maximum sharpness, the best pinhole size will be near that which makes these discs equal in extent or

$$d \simeq 2.4\frac{\lambda D}{d} \quad \text{or} \quad d = \sqrt{2.4\lambda D}.$$

The light falling on the pinhole is proportional to its area or to d^2. The area of the image of the object is proportional to D^2. Thus L, the light per unit area on the film, is proportional to

$$\frac{d^2}{D^2} = 2.4\frac{\lambda D}{D^2} = 2.4\frac{\lambda}{D}; \qquad L \propto \frac{\lambda}{D}.$$

The exposure time is reversely proportional to L or directly proportional to D.

(I-6) The mean free path is given by $\lambda \approx 1/nS$ where $n = \text{molecules/cm}^3$ and $S = \pi d^2$ = collision cross section. Here $n = (10^{-4}/760)(6.0 \times 10^{23}/22.4 \times 10^3) = 3.53 \times 10^{12}$. A good estimate for d would be 4 times Bohr radii or 2×10^{-8} cm.

Combining these quantities

$$\lambda = 1/(3.5 \times 10^{12})\pi(4 \times 10^{-16}) = 230 \text{ cm}.$$

This shows that the atoms of active H travel directly to the walls after dissociation without striking other molecules first. The "rate-limiting step" is thus the dissociation at the filament.

The rate at which molecules strike the filament is $(1/4)n\bar{v}A$, where the mean molecular speed $\bar{v} = (8RT/\pi M)^{1/2}$. Thus

$$\frac{d}{dt}(nV) = \frac{-1}{4}n\bar{v}A \quad \text{or} \quad n = n_o e^{-(\bar{v}A/4V)t}$$

where V is the constant (1-liter) volume. Since the pressure is proportional to n, the above equation becomes

$$p = p_0 e^{-(\bar{v}A/4V)t}.$$

The decline in pressure is exponential. The pressure drops by a factor ten when

$$e^{-(\bar{v}A/4V)t} = 0.1 \quad \text{or} \quad -\frac{\bar{v}At}{4V} = -\frac{1}{0.4343}$$

or

$$t = \frac{4V}{(0.4343)\bar{v}A}.$$

Using $V = 10^3 \text{ cm}^3$, $A = 0.2 \text{cm}^2$, and

$$\bar{v} = \sqrt{\frac{(8)(8.31 \times 10^7)(300)}{\pi(2)}} = 17.8 \times 10^4 \frac{\text{cm}}{\text{sec}}$$

we find

$$t = \frac{(4)(10^3)}{(.4343)(17.8 \times 10^4)(0.2)}$$

$$= 0.26 \text{ sec. per pressure drop by factor 10.}$$

For a pressure reduction by 10^3, $t = (3)(0.26) = .78$ sec.

(I-7) The P-T diagram for water is sketched below. The melting curve has a negative slope and there is a discontinuity of slope at the triple point between the sublimation and vaporization curves. Both of these effects can be understood from the Clausius-Clapyron equation:

$$\frac{dP}{dT} = \frac{L}{T\Delta V} .$$

The melting curve has negative slope for water because the volume of the solid is greater than the volume of the liquid. The slope of the vaporization curve is less than the slope of the sublimation curve because the latent heat is less for vaporization than for sublimation by the latent heat of melting.

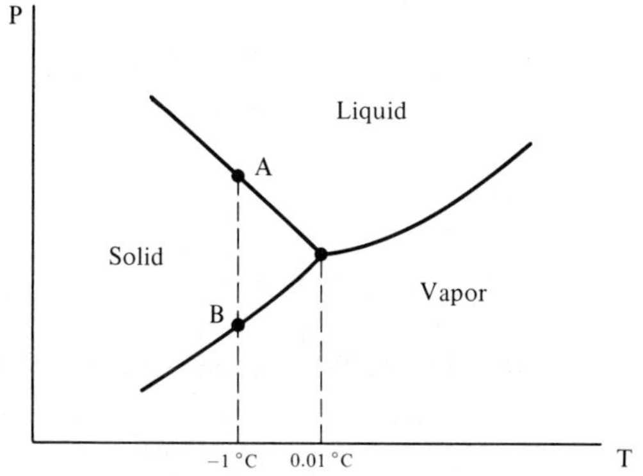

The process described in the problem is the vertical track going downward on the diagram through points A and B. At point A solidification of the high-pressure liquid occurs. At point B vaporization of the solid occurs.

To calculate P_A we must reckon the change in pressure of the melting curve for a reduction in temperature of 0.99°.

$$\frac{dP}{dT} = \frac{L}{T\Delta V} \approx \frac{(80)(4.186 \times 10^7)}{(273)(1.0001 - 1.0907)} \text{dyne/cm}^2\text{deg} \times \frac{1 \text{ atm}}{1.013 \times 10^6 \text{ dyne/cm}^2}$$

$$= -134\text{atm/deg}.$$

Thus

$$P_A = P(\text{triple point}) + (134)(0.99)$$

$$= P(\text{triple point}) + 133 \text{ atm}$$

$$\simeq 134 \text{ atm.}$$

To calculate P_B, we must reckon the change in pressure of the sublimation curve for a reduction in temperature of 0.99°. For simplicity, we will assume this vapor obeys the ideal gas equation $PV = nRT$. Therefore

$$V_{vapor} = (22.4 \times \frac{10^3}{18})(\frac{760}{4.58})$$

$$= 2.06 \times 10^5 \text{ cc/g,}$$

and

$$\frac{dP}{dT} = \frac{(80 + 596)(4.186 \times 10^7)}{(273)(2.06 \times 10^5)} \times \frac{1 \text{ mm of Hg}}{(1.359)(980)\text{dyne/cm}^3}$$

$$= 0.378 \text{ mm of Hg/deg}$$

from which we find

$$P_B = P(\text{triple point}) - (0.378)(0.99) = 4.58 - 0.38 = 4.20 \text{ mm of Hg.}$$

(II-1)

a. $n = \frac{N_o}{V_o} = \frac{6.02 \times 10^{23}}{22.4 \times 10^3} \approx 3 \times 10^{19}$ molecules/cm^3.

b. mean free path $L = 1/\pi n\sigma^2$, where σ is the diameter of the N_2 molecules. Since

$$\sigma \approx 3 \times 10^{-8} \text{ cm}$$

then

$$L = \frac{1}{\pi \times 3 \times 10^{19} \times (3 \times 10^{-8})^2} = \frac{1}{85 \times 10^3}$$

$$\approx 1.2 \times 10^{-5} \text{ cm.}$$

c. The mean kinetic energy of one N_2 molecule is

$$\frac{1}{2} m \langle v^2 \rangle$$

which should equal to $(3/2)kT$. We have

$$\frac{1}{2} m \langle v^2 \rangle = \frac{3}{2} kT \tag{1}$$

or

$$\sqrt{\langle v^2 \rangle} = \sqrt{\frac{3kT}{m}} . \tag{2}$$

The work done to the earth gravitational field when a particle escapes from the surface of the earth is

$$\text{work} = \text{potential difference} \times m = -mg \int_{r_o}^{\infty} \frac{r_o^2}{r^2} \, dr \tag{3}$$

$$= mgr_o$$

where r_o is the radius of the earth. Using (1) and (3) we find the condition for the N_2 molecules to escape from the surface of the earth is:

$$\frac{3}{2} kT = mgr_o$$

or

$$T = \frac{2mgr_o}{3k} = \frac{2Mgr_o}{3R} = \frac{2 \times 28 \times 980 \times 6.38 \times 10^8}{3 \times 8.31 \times 10^7}$$

$\sim 140,000°$K, where $M = 28$ gm is the molecular weight of N_2.

(II-2)

a. The velocity of the sound in air is

$$V = \sqrt{\frac{\gamma RT}{M}},$$

where M is the mean molecular weight of air ~ 29 gm,

$$\gamma = \frac{c_p}{c_v} = 1.40$$

$R = 8.31 \times 10^7$ ergs/mole and $T = 273°K$

from which we find

$V \sim 331$ m/sec.

b. Efficiency = $(T_1 - T_2)/T_1$ with $T_1 > T_2$.

c. The temperature of the system drops to a very low value ~ 0.01°K.

(II-3)

a. Let $g_1(P, T)$ and $g_2(P, T)$ be the Gibbs function for the two phases of the substance. In the first-order phase transition, the following properties hold:

$$g_1(P, T) = g_2(P, T)$$

but

$$\frac{\partial g_1(P, T)}{\partial T} \neq \frac{\partial g_2(P, T)}{\partial T}$$

and

$$\frac{\partial g_1(P, T)}{\partial P} \neq \frac{\partial g_2(P, T)}{\partial P}$$

where T and P are the transition temperature and vapor pressure, respectively. For a simple system which can be represented by P, v, and T we have

$dg = -sdT + vdP$

which implies

$\frac{\partial g}{\partial T} = -s$ and $\frac{\partial g}{\partial P} = v.$

Therefore we see that

$s_1 \neq s_2$ and $v_1 \neq v_2$.

b. $s = k \ln W$ where W is the probability of occurrence of the state.

c. From problem (I-1) we know the molar heat capacity is $3R$ and the specific heat is

$$c_v = \frac{3R}{63} \sim 10^{-1} \frac{\text{cal}}{\text{gm-}^\circ\text{c}}.$$

(II-4) Let $S(T)$ be the heat capacity of the object, $W_{in}(T)$ be the input power, and $W_{loss}(T)$ be the heat loss per second as functions of the temperature T. At equilibrium, we have the relation

$$W_{in}(T) = W_{loss}(T).$$

During the cooling, $W_{loss}(T)$ is related to S and T by

$$S(T)\frac{\Delta T}{\Delta t} = W_{loss}(T) \tag{1}$$

where $\Delta T/\Delta t$ is the temperature change per unit time. Using the cooling data, we first plot both T and $\Delta T/\Delta t$ as functions of t (see the following plot). From these curves we obtain the value of $\Delta T/\Delta t$ corresponding to $T = 25°, 30°, 35°$, and $40°$ C. Substituting $\Delta T/\Delta t$ into (1), we obtain $S(T)$ which is listed in the table below.

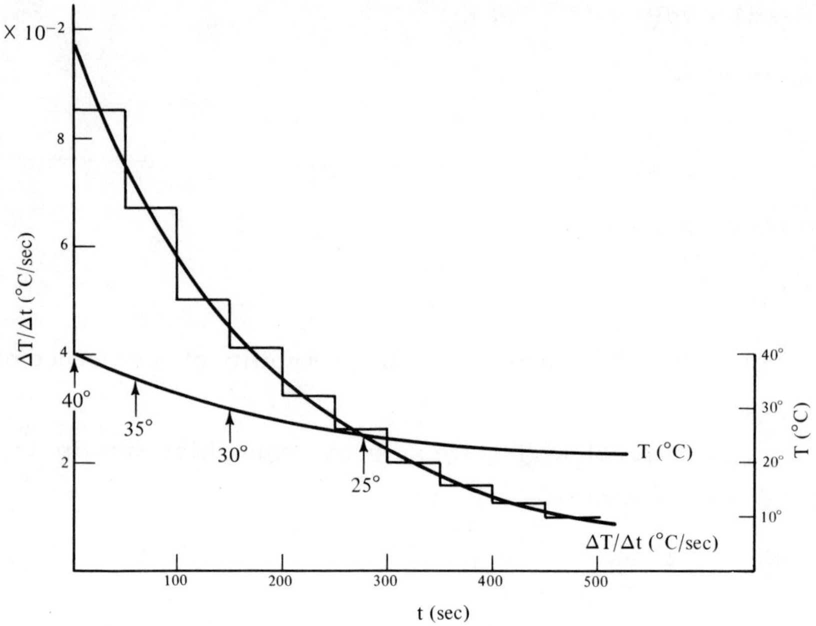

Temp.	$W_{in} = W_{loss}$	$\frac{\Delta T}{\Delta t}$	$S = \frac{W_{in}}{\Delta T/\Delta t}$ (joules/100gm-deg)
25° C	1	0.026	38
30° C	2	0.046	43.5
35° C	3	0.072	41.6
40° C	4	0.097	41

It is trivial to divide the above value of $S(T)$ by the mass of the object (100 gm) to express $s(T)$ in unit of joules/gm-deg.

(II-5) According to Boltzmann distribution we find $N_i(E_i)$ is proportional to $\exp(-E_i/kT)$ where N_i is the number of particles in state i (i = 1, 2, 3). Therefore we have the relation

$$N_1 : N_2 : N_3 = 1 : e^{-1} : e^{-2} \tag{1}$$

for $E_1 = 0$, $E_2 = E = kT$, and $E_3 = 2E$. Since the total number of particles is conserved, we find

$$N_1 + N_2 + N_3 = N. \tag{2}$$

From (1) and (2), we obtain

$$N_1 = \frac{N}{(1 + 1/e + 1/e^2)},$$

$$N_2 = \frac{N}{(1 + 1/e + 1/e^2)} \frac{1}{e}$$

and

$$N_3 = \frac{N}{(1 + 1/e + 1/e^2)} \frac{1}{e^2}. \tag{3}$$

The total energy of the system is

$$E = E_1N_1 + E_2N_2 + E_3N_3$$

$$= \frac{NkT}{1 + 1/e + 1/e^2} (\frac{1}{e} + 2/e^2)$$

which is known to be 1000 kT. Therefore we have

$$N = \frac{1000(1 + e^{-1} + e^{-2})}{e^{-1} + 2e^{-2}} \approx 2400.$$

(II-6)

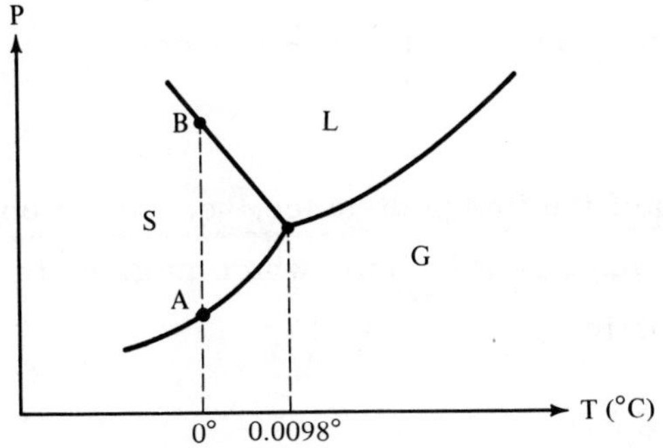

Use Gibbs' phase rule

$$F = c - p + 2 \tag{1}$$

where F is the number of degrees of freedom, c is the number of components in the system and p is the number of phases.

a. In this problem, the temperature is fixed at 0° C. We have lost one degree of freedom. With $c = 1$, $p = 2$, and fixed temperature, we find $F = 1 - 1 = 0$. This means that the equilibrium state is just one point in the p-T diagram which is determined by the pressure only, and in that state there cannot be more than two phases present. The initial vapor pressure is zero. From the above pressure-temperature diagram, we see both water and ice have to vaporize. The vapor pressure is thus increasing until it reaches the point A. The final state is that all the water has vanished; part of it changes into vapor and the rest of it becomes ice. The final state of the system is at A in the p-T diagram.

b. If the temperature is not kept constant, from the phase rule we find $p = 3$ for $c = 1$ and $F = 0$, i.e., all three phases can exist simultaneously. The final state is the triple point.

(II-7) The minimum angle of resolution for a circular lens is

$$\theta \equiv \frac{x}{D} = 1.22\frac{\lambda}{d}; \quad \text{or} \quad d = 1.22\frac{\lambda D}{x}.$$

Using $\lambda = 500$ Å, $x = 1$ ft, and $D = 100{,}000$ ft, we get

$$d = 1.22 \times 5 \times 10^{-5}\ \text{cm} \times 10^{5}\ \text{ft}/1\ \text{ft} \simeq 6.1\ \text{cm}.$$

(II-8) Closer.

(II-9) The ratio of the image distance, q, to the object distance, p, equals the magnification factor which is given to be 24. Therefore we obtain

$$p = \frac{q}{24} = \frac{5}{8}\ \text{ft}.$$

Substituting the known values of p and q into the lens formula

$$\frac{1}{f} = \frac{1}{p} + \frac{1}{q},$$

we find

$$f = \frac{3}{5} \text{ ft.}$$

(II-10) Usually it is 25 cm from the eye; this is taken to be the standard distance of most distinct vision.

(II-11) The amplitude at the center is proportional to the width of the slit. Therefore the intensity of the center of the pattern increases by a factor of 4. The total transmitted energy increases by a factor of 2.

(II-12) Since the reflectance of glass for normal incidence is small, the intensity of the reflected fringe system is less intense than that of the transmitted fringe system.

(II-13) Linear polarization.

(II-14) The intensity goes to zero when the radius of the opening corresponds to the radius of the second Fresnel zone. We have

$$R = \sqrt{2\lambda D}.$$

(II-15) Using $1/p + 1/(\ell - p) = 1/f$, we can solve for p:

$$p = \frac{\ell}{2}\left(1 \pm \sqrt{1 - \frac{4f}{\ell}}\right).$$

The conditions are:

$\ell < 4f$; $\ell = 4f$; and $\ell > 4f$

corresponding to 0, 1, and 2 real values of p, respectively.

(II-16) $n = 1.5$ and $\dfrac{dn}{dx} = 10^{-5}/\text{Å}$.

(II-17) $p_1 + q_1 = \ell$ where $M_1 = q_1/p_1 = 1/2$; thus $2q_1 + q_1 = \ell$. Therefore $p_1 = \frac{2}{3}\ell$, $q_1 = \frac{1}{3}\ell$. If we move so that $p_2 = \frac{1}{3}\ell$, then $q_2 = \frac{2}{3}\ell$ and $M_2 = q_2/p_2 = 2$. Thus $M_2 \times 1$ inch = 2 inch.

(II-18) The first minimum in the Fraunhofer diffraction pattern appears at

$$\theta = 1.22\frac{\lambda}{d}.$$

Therefore the diameter of the image of the star is

$$d_s = 2 \times (1.22\frac{\lambda}{d})D = 2.44\frac{\lambda D}{d}$$

where d is the diameter of the pupil and D is the image distance. If we assume that $d \simeq 5$ mm = 0.5 cm and that $D \simeq 3.0$ cm so that $D/d \simeq 6$, then

$$d_s = 2.44 \times 6\lambda = 14.6 \times 5 \times 10^{-5}\,\text{cm} \simeq 7 \times 10^{-4}\,\text{cm}.$$

(II-19) For a virtual object behind the mirror.

(II-20) According to the Rayleigh criterion the maximum resolving power of an optical device corresponds to the condition that the principal maximum of a diffraction pattern of one point object falls exactly on the first minimum of the diffraction pattern of an adjacent point object. This is taken to correspond to the smallest angular separation the device can resolve. For the circular opening we find

$$\theta = 1.22\frac{\lambda}{d}$$

where d is the diameter of the aperture.

(II-21) When $f < \ell < 2f$, the focal length of the system is negative.

(III-1)

a. By Snell's law we have

$$\sin\theta_1 = n \sin\theta_2. \tag{1}$$

For small θ_1 and α, (1) becomes

$$\theta_1 = n\theta_2. \tag{2}$$

Similarly,

$$\theta_4 = n\theta_3. \tag{3}$$

Using simple geometrical relations between angles of triangles, we obtain

$$\alpha + \pi - a = \pi \quad \text{or} \quad a = \alpha$$

and

$$\alpha = (\theta_2 + \theta_3) \tag{4}$$

and

$$\delta = (\theta_1 - \theta_2) + (\theta_4 - \theta_3). \tag{5}$$

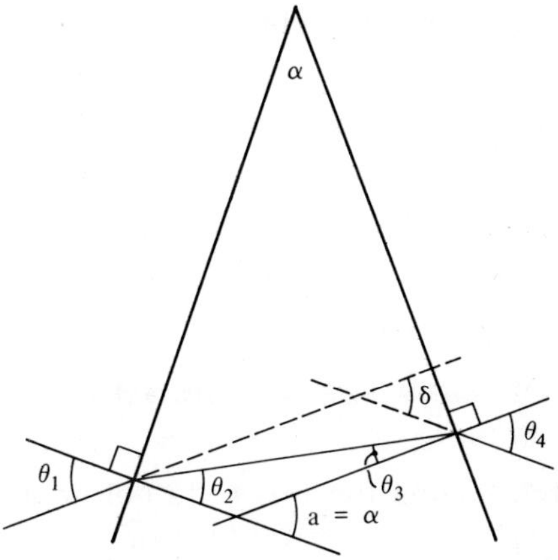

From (2), (3), (4), and (5) we get

$$\delta = (n - 1)\alpha,$$

which is independent of incident angle θ_1.

b. Let o' be the virtual position of the object viewing behind the prism. From part a, we see each ray is deviated by an angle δ after passing through the prism.

From the attached figure we find that to the first order in α and β, the object distance does not change when the object moves from o to o'. Substituting

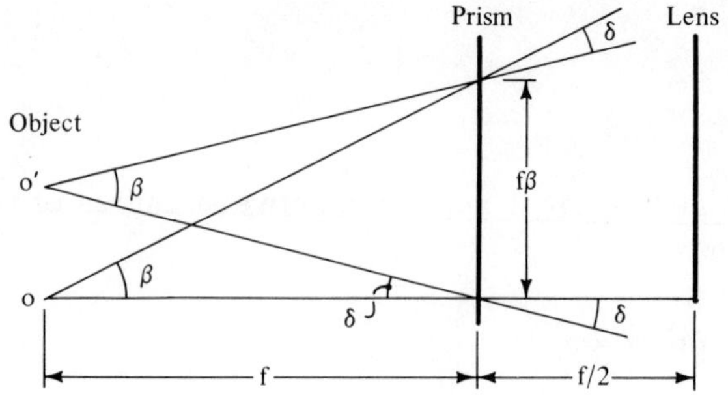

$$p = f + \frac{f}{2} = \frac{3}{2}f$$

into the Gaussian lens formula, we obtain

$$\frac{1}{(3/2)f} + \frac{1}{q} = \frac{1}{f} \tag{6}$$

or the image distance along the axis

$$q = 3f. \tag{7}$$

The transverse object distance oo' is related to the angle δ by $oo' = f\delta$. The image transverse distance II' is simply related to oo' by

$$-\frac{II'}{oo'} = \frac{q}{p} \quad \text{or} \quad II' = -2\frac{f}{f}oo' = -2f\delta = -2f(n-1)\alpha$$

where the minus sign means the image is below the axis.

(III-2) The device is called a transmission echelon, which is similar to a grating. Let the phase change from one "slit" to the next be δ. We find

$$\delta = \frac{2\pi((n - \cos\theta)t + s\sin\theta)}{\lambda}.$$

For small θ, this becomes

$$\delta \simeq \frac{2\pi((n-1)t + s\theta)}{\lambda}. \tag{1}$$

where θ is the angle of the refracted ray with respect to the incident beam. Using the formula for the intensity of an N-slit grating, we obtain

$$I \sim \frac{\sin^2 \beta}{\beta^2} \frac{\sin^2 N\frac{\delta}{2}}{\sin^2 \frac{\delta}{2}} \tag{2}$$

where

$$\beta = \frac{\pi}{\lambda} s \sin\theta \simeq \frac{\pi}{\lambda} s\theta .$$

The factor

$$\frac{\sin^2 \beta}{\beta^2}$$

is due to the single "slit" diffraction, while the factor

$$\frac{\sin^2 (N\frac{\delta}{2})}{\sin^2 \frac{\delta}{2}}$$

represents the interference term for N "slits." When

$\frac{\delta}{2} = 0, \pi, 2\pi, \ldots,$ we have constructive interference, i.e.,

$$\frac{\pi((n-1)t + s\theta)}{\lambda} = m\pi, \qquad m = 0, 1, 2, \ldots$$

or

$$\theta = \frac{1}{s}(m\lambda - (n-1)t) \equiv \theta_m \quad \text{maxima} \tag{3}$$

Substituting

$\theta \approx 0; \quad n = 1.5; \quad t = 0.5$ cm; and $\lambda = 5000$ Å

into (3) we have

$$\frac{1}{s}(m \times 5 \times 10^{-5} - (0.5)^2) \approx 0$$

from which we find the order of interference is

$$m = 5 \times 10^3. \tag{4}$$

For small $dn/d\lambda$, the angular dispersion $\Delta\theta/\Delta\lambda$ can be obtained directly from (3)

$$\frac{\Delta\theta}{\Delta\lambda} = \frac{1}{s}(m - t\frac{dn}{d\lambda})$$

$$\approx \frac{m}{s} \tag{5}$$

The resolving power depends on the angular distance, $\Delta\theta$, between the mth order of principal maximum and its nearest minimum. We find when

$$\frac{\delta}{2} = (m\pi + \frac{\pi}{N})$$

the intensity becomes zero, using (1) we get

$$\theta'_m = \frac{1}{s}(\frac{\lambda}{N} + m\lambda - (n - 1)t) \quad \text{minimum} \tag{6}$$

Using (6) and (3) we obtain the angular half width of the principal maximum

$$\Delta\theta = (\theta'_m - \theta_m) = \frac{1}{s}\frac{\lambda}{N}. \tag{7}$$

Let θ_1, θ_2 be the angular positions corresponding to the principal maxima of the waves of λ_1 and λ_2. When θ_1 and θ_2 are separated by the angle $\Delta\theta$ defined in (7), the two images are barely resolved. Therefore we have

$$\Delta\theta \sim (\theta_1 - \theta_2) = \frac{1}{s}\frac{\lambda}{N}.$$

Using (3) we get

$$\frac{1}{s}(m\lambda_1 - (n - 1)t) - \frac{1}{s}(m\lambda_2 - (n - 1)t) = \frac{1}{s}\frac{\lambda}{N}$$

from which we obtain the resolving power

$$\frac{\lambda_1}{\lambda_1 - \lambda_2} = \frac{\lambda}{\Delta\lambda} = mN$$

$$\approx 40 \times 5 \times 10^3 = 2 \times 10^5.$$

(III-3) The electric vector components for right circularly polarized light are

$$E_x = E_o \sin(\omega t - \omega \frac{z}{v_x})$$

$$E_y = E_o \sin(\omega t + 90° - \omega \frac{z}{v_y})$$

n_e is smaller than n_o, so v_e is larger than v_o. Let us define the optic axis of the first plate to be the X-axis. After the light passed through the first plate, the electric vector is

$$E_x^{(1)} \propto \sin \omega t$$

$$E_y^{(1)} \propto \sin(\omega t + 90° + 90°) = -\sin \omega t$$

which represents linear polarized light. The polarization vector makes 45° with the optic axis of the plate.

a. For $\theta = 0°$, the electric vector after passing through the second plate is

$$E_x^{(2)} \propto \sin \omega t$$

$$E_y^{(2)} \propto \sin(\omega t - 90°) = -\cos \omega t$$

It is left circularly polarized.

b. For $\theta = 45°$, there is no relative change in the phases of E_x and E_y. Therefore it is still linearly polarized just as when it emerged from the first plate.

c. For $\theta = 90°$

$$E_x^{(2)} \propto \sin(\omega t + 90°) = \cos \omega t$$

$E_y^{(2)} \propto -\sin \omega t$

It is right circularly polarized.

d. Two quarter-wave plates are identical to a single half-wave plate only if $\theta = 0$.

(III-4)

a. Let θ be the angle between the axes of the two polarizers. We can identify the two polarizers by the fact that, for an unpolarized light source, the intensity of light after passing through the two devices is proportional to $\cos^2 \theta$.

b. Set the two linear polarizers in such a way that $\theta = 90°$. Light is totally attenuated by the system in this case. Then:

c. Put one of the remaining three devices between the two linear polarizers, and rotate the device around the light direction. Let the intensity of the light after the first polarizer be $I/2$. If

i. The maximum intensity as seen from behind the second linear polarizer is $I/2$; then the device is a half-wave plate. (This corresponds to the case that the half-wave plate makes 45° with either of the two polarizers.)

ii. The intensity after L-P-2 is constant and equals to $I/8$; then it is a circular polarizer (since the intensity decreases by a factor of 2 at each of the three devices).

iii. The intensity after L-P-2 varies between zero and $I/4$; then it is a quarter-wave plate. (We get maximum intensity when the quarter-wave plate makes 45° with either of the two polarizers. Light is attenuated by a factor of 2 at each of the polarizers.)

d. If there were only one linear polarizer, we can separate the linear polarizer and the circular polarizer from the other two based on the property that either of the two polarizers reduces the intensity of the lamp light by a factor of 2 while the others do not.

Furthermore we can tell which is which by playing around with any one of the two plates put between the two polarizers until one of the following phenomena is observed:

i. The system of one circular polarizer in front of a quarter-wave plate can serve as a linear polarizer, i.e., as shown in the figure below:

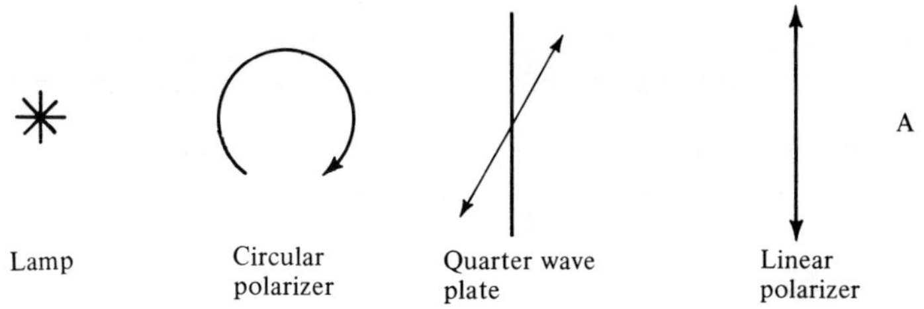

The intensity at A varies as $\cos^2(\theta + \frac{\pi}{4})$ when we rotate the linear polarizer. θ is the angle between the axis of the linear polarizer and that of the quarter-wave plate. Note that the intensity at A does not change as we rotate the first (circular) polarizer.

ii. The system of a linear polarizer in front of a quarter-wave plate serves as a circular polarizer when the quarter-wave plate makes $\pm 45°$ with respect to the axis of the linear polarizer, i.e.,

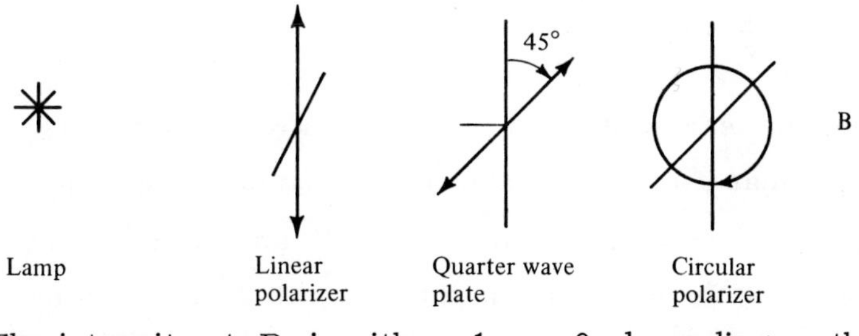

The intensity at B is either 1 or 0 depending on the relative position of the linear polarizer and the plate, i.e., whether the angle between them is + 45° or -45°. Note that the intensity at B is constant as we rotate the second (the circular) polarizer.

The system of a half-wave plate and any two devices of the rest cannot totally attenuate an initially unpolarized light. Therefore when we observe i and ii, we have identified these four devices.

(III-5)

a. Entropy is defined in statistical mechanics by

$$S = k \ln W(n_i)$$

where n_i is the average number of particles in the ith state and $W(n_i)$ is the number of ways the system can be produced corresponding to the set of n_i, $i = 1, 2, \ldots$. Here

$$W(n_i) = \frac{N!}{n_1! n_2!}$$

where n_1 and n_2 are the number of atoms with spin up and down, respectively; $n_1 + n_2 = N$. Using Stirling's approximation

$$\ln n! = n(\ln n - 1) \quad \text{for} \quad n \gg 1$$

we obtain

$$S = k[N \ln N - (n_1 \ln n_1 + n_2 \ln n_2) - N + (n_1 + n_2)]$$

$$= -k \sum_i n_i \ln \frac{n_i}{N} \qquad i = 1, 2$$

$$= -k(n_1 \ln \frac{n_1}{N} + n_2 \ln \frac{n_2}{N}) \tag{1}$$

Entropy may be defined in another way: it is found by experiment that if the heat added at each point of the path on any diagram is divided by the temperature T and the resulting ratio is integrated over the entire path, the integrated ratio is a constant. Since conserved quantities are of vital interest in physics, we define

$$S_T = \int_0^T \frac{dQ}{T} = \int_0^T \frac{C dT}{T}. \tag{2}$$

b. As $T \to 0$, all the spins are aligned. Therefore we have $n_1 = N$, $n_2 = 0$, and from (1) we find $S_o = 0$. As $T \to \infty$, the spins are randomly oriented, i.e., $n_1 = n_2$. We have

$$S_\infty = \frac{-kN \ln n_1}{N} = kN \ln 2. \tag{3}$$

Using (2) and (3) we obtain

$$S_\infty \equiv \int_0^\infty \frac{C(T)dT}{T} = kN \ln 2.$$

(III-6) Since the temperature of the vessel is kept constant, the mean velocity of the gas molecules, v, is constant. The projected mean velocity of the molecules is

$$\bar{v} = \frac{\int_0^1 n(v \cos\theta)d\cos\theta}{\int_{-1}^{+1} nd\cos\theta} = \frac{v}{4}.$$

The number of molecules which pass through the hole with area A per unit time is

$$-\frac{dn}{dt} = \frac{n\bar{v}A}{V} = \frac{nvA}{4V} \tag{1}$$

where n is the number of molecules left in the vessel and V is the volume of the vessel. The solution of (1) is

$$n = n_o \exp -(\frac{vA}{4V}t).$$

From which we find for $n = \frac{n_o}{e}$,

$$t = \frac{4V}{vA}, \quad \text{where } v = \sqrt{\frac{8kT}{\pi m}}.$$

(III-7) According to the Stefan-Boltzmann law, the radiancy of a black body at temperature T is equal to

$E(T) = kT^4$

where k is called Botzmann constant. Using the inverse square law for radiation energy we find that the intensity of radiation from the sun on the earth is

$$E_s = kT_o^4\left(\frac{R}{D}\right)^2.$$

The total radiation energy the earth receives is

$$E_e = kT_o^4\left(\frac{R}{D}\right)^2 \pi r^2 \quad (1)$$

which is assumed to be equal to the total energy the earth radiates

$$E_e = kT^4 4\pi r^2. \quad (2)$$

From (1) and (2) we find

$$T^4 = T_o^4\left(\frac{R}{2D}\right)^2 \quad \text{or} \quad T = \sqrt{\frac{R}{2D}}\, T_o = 4.82 \times 10^{-2} T_o$$

or

$T \sim 275°K$

which is about room temperature.

(IV-1) From the first law of thermodynamics and the definition of entropy we have

$$TdS = C_v dT + PdV. \quad (1)$$

a. The entropy change of the N_2 gas when temperature changes from 0° C to 100° C at constant volume is,

$$\Delta S_{N_2} = \int_{T_i}^{T_f} \frac{C_v dT}{T} = C_v \ln \frac{373}{273} = 0.312\, C_v \quad (2)$$

The specific heat for diatomic gases at room temperature is (5/2)R per mole, or for one liter of gas

$$C_v = \frac{5}{2} R \frac{1}{22.4} \simeq \frac{5}{2} \times 2 \times \frac{1}{22.4} = 0.223 \frac{\text{cal}}{\text{deg-liter}}$$

Substituting the value of C_v into (2) we get

$$\Delta S_{N_2} = 0.312 \times 0.223 = 0.0696 \text{ cal/K.}$$

The entropy change of the reservoir at 100° C is

$$\Delta S_{\text{reservoir}} = \frac{\Delta Q}{T} = \frac{C_v \Delta T}{T} = \frac{0.223 \times (-100)}{373} = -0.0598 \frac{\text{cal}}{\text{K}}$$

The change in entropy of the universe is of

$$\Delta S_{\text{Univ}} = \Delta S_{N_2} + \Delta S_{\text{reservoir}} = 8 \times 10^{-4} \frac{\text{cal}}{\text{K}}.$$

b. If one wall of this cylinder is movable, we can first compress the gas adiabatically until the temperature of the gas increases to 100° C; then we bring it in contact with the reservoir and let the gas expand isothermally back to its original volume.

There is no change in entropy when the process is adiabatic. Furthermore, since the change in entropy of the N_2 gas equals Q/T, while that of the reservoir is $-Q/T$ when the process is isothermal, the total entropy change of the world is zero. $0 + Q/T + (-Q/T) = 0$.

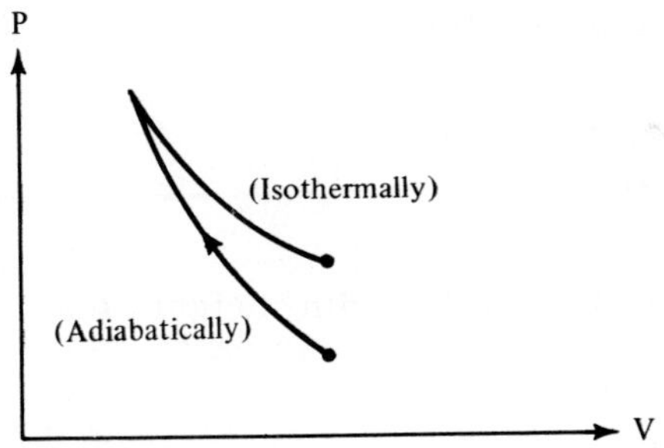

(IV-2)

Substituting $p = 20$ and $f = 10$ into the lens formula

$$\frac{1}{p} + \frac{1}{q} = \frac{1}{f} \tag{1}$$

we find the first image distance $q = 20$ cm. The object distance for the second lens is $70 - 20 = 50$ cm. The final image distance is then $50/4$ cm.

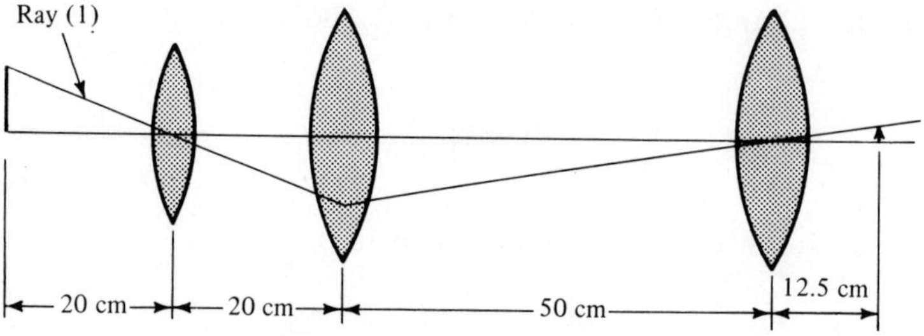

If a third lens is put exactly at the position of the first image, the object distance of the second lens is unchanged, and the position of the final image is unchanged. In order to increase the collecting efficiency, f should be positive.

In the above figure, ray (1) would be lost if the third lens were not inserted. The third lens is sometimes called "field lens" in beam optics.

(IV-3) Substituting

$R = 50$ cm, $r_3 = 0.09$ cm and $r_{23} = 0.25$ cm

into the formula for bright fringes

$$2\left(\frac{r_m^2}{2R} + d\right) = \left(m + \frac{1}{2}\right)\lambda$$

where r_m is the radius of the mth bright ring. For $m = 3$, we obtain

$$2d + \frac{0.0081}{50} = \frac{7}{2}\lambda \tag{1}$$

and for $m = 23$, we obtain

$$2d + \frac{0.0625}{50} = \frac{47}{2}\lambda \tag{2}$$

where d is some unknown constant which exists because the plates do not necessarily touch at the point of closest approach. Eliminating d between (1), (2), we obtain

$$\lambda = 5.44 \times 10^{-5}\ \text{cm} = 5440\,\text{Å}.$$

(IV-4)

a. Since both state 1 and state 2 are equilibrium states between water and vapor, at constant temperature T_1, the vapor pressure of water, p_w, which is generally a function of T only, remains constant. Let the partial pressure of the air in state 1 be p_o. We have the following relations,

$$p_1 = p_w + p_o = 3 \tag{1}$$

$$p_2 = p_w + \frac{p_o}{2} = 2 \tag{2}$$

from which we find

$$p_w = 1 \text{ and } p_o = 2 \quad \text{atmospheres.}$$

Since p_w equals one atmosphere, the water would be boiling if the total pressure were 1 atmosphere. Therefore we know

$$T_1 = 100°\,C.$$

b. If $V_3 = 2V_2$, $p_3 = p_2V_2/V_3 = 1$ atmosphere.

c. Using $V_2 = 44.8$ liters, $p_w = 1$ atmosphere, and $T_1 = 100°$ C, we find

$$n_w = n_{air} = \frac{273}{373} \times 2 = 1.46 \text{ mole.}$$

The molecular weight of H_2O is 18 gm while that of air is 29 gm. Therefore the mass of water $= 18 \times n_w = 26.28$ gm and the mass of air $= 29 \times n_{air} = 42.34$ gm.

(IV-5)

a. The change in entropy is zero because the internal energy of an ideal gas is a function of temperature only. For free expansion of an ideal gas we have

$$T\Delta S = \Delta Q = \Delta U + \Delta W = 0 + 0 = 0.$$

b. The entropy of one mole of ideal gas at given T and p is

$$S(T, p) = C_p \ln T - R \ln p + K$$

where K is a numerical constant. The entropy of a mixture of two ideal gases of one mole each, starting with equal volumes and temperature, is

$$S_{12}(T, p) = (C_{p1} + C_{p2}) \ln T - 2R \ln p' + 2K$$

where p' is the partial pressure of either of the gases. Since the volume occupied by each gas has doubled, we have

$$p' = \frac{1}{2} p.$$

Therefore the entropy change is

$$\begin{aligned} S_{12} - 2S &= 2R \ln p - 2R \ln \frac{p}{2} \\ &= 2R \ln 2 \end{aligned}$$

c. In the case of mixture of same gas p' in the final state equals p in the initial state; therefore there is no entropy change.

(IV-6)

a. The energy of each photon of wavelength 6200 Å is

$$E_\gamma = \frac{2\pi\hbar c}{\lambda} \sim 2\text{eV}. \tag{1}$$

The total number of photons in 1 watt of light is

$$n_\gamma = \frac{1}{2 \times 1.6 \times 10^{-19}} = 3.1 \times 10^{18}/\text{sec} \tag{2}$$

Since the angular momentum of a single photon is

$$\frac{E_\gamma}{\omega} = \frac{\hbar\omega}{\omega} = \hbar, \tag{3}$$

the total angular momentum transferred per second is

$$\tau = \frac{dJ}{dt} = n_\gamma \hbar \sim 3 \times 10^{18} \times 10^{-27} \sim 3 \times 10^{-9} \text{ dyne-cm} \tag{4}$$

The direction is parallel or antiparallel to the direction of propagation depending on whether the photon is L.P. or R.P.

b. Let R(L).P. stand for a right (left)-handed circularly polarized photon. The relative phase change of the electric vectors of light externally reflected from a denser material is π. Therefore a R.P. photon before reflection becomes L.P. after reflection. But the direction of the angular momentum in space has not changed. Therefore the total angular momentum transferred per second is zero.

c. A R.P. photon becomes L.P. after passing through a half-wave plate. The angular momentum transferred is twice as much as that in a.

$$\tau_c = 2\tau \tag{5}$$

d. A R.P. photon becomes L.P. after the half-wave plate. After reflection the L.P. photon is R.P. This photon becomes L.P. after passing back through the half-wave plate. Considering only this final photon and the initial photon, we find the angular momentum transferred is zero.

e.

i. R.P. photons become L.P. after the half-wave plate.

ii. L.P. photons become linearly polarized after the quarter-wave plate.

iii. Linear polarized photons are still linearly polarized but, after reflection from the silvered surface, the polarization direction is opposite to the original direction.

iv. Linear polarized photons become L. P. polarized after passing back through the quarter-wave plate.

v. L. P. photons become R. P. after passing back through the half-wave plate.

As far as the slab is concerned, it receives one R. P. photon from below (the angular momentum transferred is $\hbar\downarrow$) and one L. P. photon from above ($\downarrow$) and it emits one L. P. photon upward ($\uparrow$) and one R. P. photon downward ($\uparrow$). So we have

$$\downarrow + \downarrow - \uparrow - \uparrow = 4\downarrow$$

$$\tau_e = 4\tau.$$

(IV-7)

a. The number of boys born is

$$n_B \propto P + (1 - P)P + (1 - P)^2 P + \dots + (1 - P)^n P + \dots$$

$$= P(1 + (1 - P) + (1 - P)^2 + \dots)$$

where P is the probability that a born child turns out to be a boy, $P = 51\%$. The number of girls is

$$n_g \propto (1 - P) + (1 - P)^2 + \dots + (1 - P)^n + \dots$$

$$= (1 - P)\,(1 + (1 - P) + (1 - P)^2 + \dots)$$

Therefore the ratio of n_B and n_g is still $P/(1 - P)$. 51% of the children born are boys.

b. Let the apparatus

i. Emit electromagnetic waves at a known frequency ν_o toward the object.

ii. Detect electromagnetic waves at several different frequencies ν_o, $\nu_o \pm \Delta\nu$, $\nu_o \pm 2\Delta\nu$ ... from the object.

iii. Plot the intensity of the detected waves against frequency. If the object is a black body, the intensity is a smooth function of frequency. If the object is not a black body there will be a spike at the frequency of ν_0. This is so because the frequency of the reflected wave is the same as that of the incoming wave.

iv. As a check, we can change the frequency of the emitted wave to $(\nu_0 + \Delta\nu)$. If the object is a black body the detected spectrum is the same as it was before. If the object is not a black body, the peak of the spectrum will move to $(\nu_0 + \Delta\nu)$. This is the principle of radar.

(IV-8) Since the protons are at thermal equilibrium, the distribution of the states is simply the Maxwell-Boltzmann distribution function:

$$\frac{N(\text{up})}{N(\text{down})} \propto \exp - \frac{((E(\text{up}) - E(\text{down}))}{kT}$$

$$= \exp \frac{\hbar\omega}{kT} = \exp \frac{N\hbar\omega}{RT}$$

$$= \exp \frac{6 \times 10^{23} \times 10^{-34} \times 10^{8}}{2 \times 4.18 \times 300}$$

$$= K \approx 1 + 2.4 \times 10^{-6}.$$

Therefore the polarization P is

$$P = \frac{N(\text{up}) - N(\text{down})}{N(\text{up}) + N(\text{down})} = \frac{K - 1}{K + 1}$$

$$= 1.2 \times 10^{-6}.$$

(V-1)

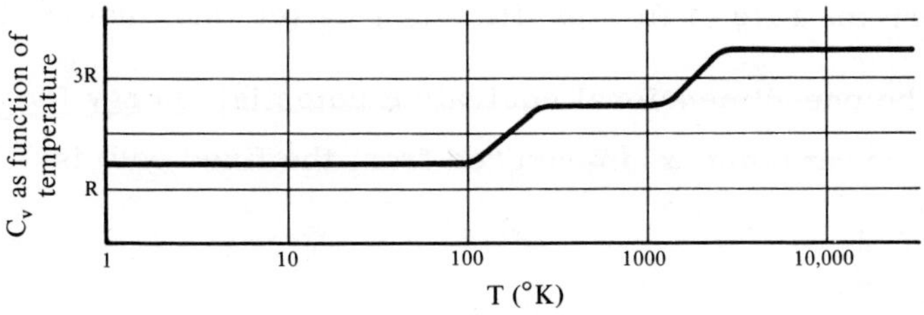

At low temperature we find the total energy E and heat capacity c_v are

$$E = \frac{3}{2}RT \text{ and } c_v = \frac{3}{2}R$$

corresponding to the three-component momenta that contribute quadratic terms to the kinetic energy of the molecule. The vibrational and rotational motions are forbidden because the thermal energy kT is less than the energy difference between the first excited state and the ground state. As temperature increases, rotational motion occurs. We obtain

$$c_v = \frac{5}{2}R$$

corresponding to the two additional rotational energy terms that are now present. At higher temperature we find $c_v = (7/2)R$ corresponding to the two additional vibrational energy terms. At very high temperature we have to further consider the excitation energy of the electrons.

(V-2)

$$\dot{S} = \frac{\dot{Q}}{T} = \frac{RI^2}{T} = \frac{100I^2}{303} = 0.33I^2 \text{ joules/}^\circ\text{K/sec.}$$

(V-3) Smaller. The amount of heat exchanged is proportional to rate of flow of the material. Below the λ point ($T \approx 2.2^\circ K$), He^4 becomes superfluid and the rate of flow is immense. He^3 behaves as a normal fluid with normal viscosity at low temperature. Therefore the rate of flow of He^3 is smaller than that of He^4.

(V-4) The one-dimensional equivalent potential energy for a molecule of mass m at point x (measured from the fixed end) is

$$E = \frac{-m\omega^2 x^2}{2}.$$

According to Boltzmann distribution, the density $\rho(x)$ is

$$\rho(x) = \rho(o) \exp\left(\frac{M\omega^2 x^2}{2RT}\right)$$

where M is the molar weight of the gas and R is the universal gas constant. Note: $M = N_o m$ and $R = N_o k$, where N_o is Avogadro's number and k is the Boltzmann constant.

(V-5) According to the equipartition principle we have

$$\overline{P.E.} = \frac{1}{2}C\overline{\theta^2} = \frac{1}{2}kT \text{ and } \overline{K.E.} = \frac{1}{2}I\overline{\dot{\theta}^2} = \frac{1}{2}kT,$$

therefore

$$\overline{\theta^2} = \frac{kT}{C} \text{ and } \overline{\dot{\theta}^2} = \frac{kT}{I}.$$

which are independent of pressure.

(V-6) The heat absorbed by the system in the process $B \to C$ is

$$Q_{BC} = \int_B^C Tds = \frac{3}{2}(5000) = 75,000 \text{ joules}.$$

The heat emitted during the process $C \to A$ is (-100×500) or -50,000 joules. Since Q_{AB} is zero we obtain the efficiency of the engine

$$\eta = \frac{\text{work}}{Q_{BC}} = \frac{Q_{BC} + Q_{CA}}{Q_{BC}} = 33.3\%.$$

(V-7) The difference in specific enthalpy between two neighboring equilibrium states is

$$dh = \left(\frac{\partial h}{\partial T}\right)_P dT + \left(\frac{\partial h}{\partial P}\right)_T dP$$

where

$$\left(\frac{\partial h}{\partial T}\right)_P$$

is defined to be c_P.

If the gas is to be cooled by decreasing P, we find

$$\left(\frac{\partial T}{\partial P}\right)_h > 0 . \tag{1}$$

Using the relation

$$\left.\frac{\partial h}{\partial P}\right|_T = - \frac{(\partial h/\partial T)_P}{(\partial P/\partial T)_h} = - \frac{c_P}{(\partial P/\partial T)_h}$$

and the fact that $c_P > 0$, we obtain

$$\left.\frac{\partial h}{\partial P}\right|_T = - \frac{c_P}{(\partial P/\partial T)_h} < 0.$$

(V-8) The coefficient of viscosity is essentially independent of the pressure. Therefore the mean-square displacement is the same. (At constant temperature, the viscosity increases some 20% to 40% for a pressure increase of 1000 atmospheres.)

(V-9) Using the first TdS equation for black-body radiation,

$$TdS = C_V dT + T\left(\frac{\partial P}{\partial T}\right)_V dV$$

and the relations

$$P = \frac{u}{3} = \frac{4\sigma T^4}{3c} \quad \text{and} \quad C_V = \frac{\partial U}{\partial T} = \frac{16\sigma}{c} T^3 V$$

we get

$$TdS = 0 = V\frac{16\sigma}{c}T^3 dT + T\frac{16\sigma T^3}{3c} dV$$

or

$$\frac{dV}{V} + \frac{3dT}{T} = 0$$

and after integration

VT^3 = constant so that $T_f = (\frac{V_i}{V_f})^{1/3} T_i$.

Therefore

$$T_f = \frac{1}{2} T_i \quad \text{for} \quad V_f = 8V_i.$$

(V-10)

$$\Delta S = \frac{\text{mechanical energy}}{T} = \frac{kA^2}{2T},$$

where k is the spring constant and T is the absolute temperature.

(V-11) At zero °K, all the electrons are at the lowest states. The number of states is proportional to $p^2 dp$ or $\sqrt{E}\, dE$. After integration from zero to E_{max}, we find

$$N \propto E_{max}^{3/2} \quad \text{or} \quad E_{max} \propto N^{2/3}$$

where E_{max} is Fermi energy;

$$E_{max} = \frac{\hbar^2}{2m} \left(\frac{3\pi^2 N}{V}\right)^{2/3} .$$

(V-12) Let T and T_r be the temperature of the water and the reservoir, respectively. The highest efficiency we can get between two sources of heat of temperature T and T_r is

$$\eta = \frac{\Delta W}{\Delta Q} = \frac{T_r - T}{T}$$

from which we find the minimum work needed is

$$W = -\int_{293}^{273} \frac{T_r - T}{T} C_v dT + \frac{T_r - 273}{273} (80)(1000)$$

$$= 293 \times 10^3 \ln \frac{293}{273} - 20 \times 10^3 + \frac{20 \times 80 \times 1000}{273}$$

$$\sim 7 \times 10^3 \text{ joules.}$$

(V-13) The atmospheric pressure as a function of height is

$$P = P_o \exp\left(-\frac{mgh}{kT}\right) = P_o \exp - \left(\frac{Mgh}{RT}\right).$$

Using $h = 1000$ ft $= 3 \times 10^2$ m, $g = 9.80$ m/sec^2, $M = 29$ gm/mole, $R = 8.3$ joules/mole and $T \sim 300°$K, we find

$$P \sim P_o \exp - \left(\frac{Mgh}{RT}\right) = P_o \exp(-0.034)P_o \simeq (1 - 0.034)P_o,$$

or $\Delta P = -0.034\, P_o$. Since evaporation is a first-order transition which obeys Clapeyron's equation

$$\frac{dP}{dT} = \frac{L}{T(v - v_w)} \tag{1}$$

where L is the latent heat; v and v_w are the specific volumes of the vapor and the water, respectively. Note that v_w is negligible compared with v. Using the ideal gas equation $v = RT/18P$ we obtain

$$\frac{\Delta P}{P} = \frac{18L}{RT^2}\Delta T$$

for one gram of vapor, or

$$\Delta T = \frac{2 \times 373^2}{540 \times 18}\frac{\Delta P}{P_o} \sim -1° \text{C}.$$

(V-14) Using the second Tds equation

$$Tds = c_p dT - T\left(\frac{\partial V}{\partial T}\right)_p dP$$

we obtain the change in entropy at constant pressure

$$ds = \frac{c_p dT}{T}.$$

After integration we find the entropy at $-10°$ C is

$$s - s_o = 4222\, \ell n \frac{263}{273} + 226 \quad \text{for water} \tag{1}$$

and

$$s' - s'_o = 2112\, \ell n \frac{263}{273} - 75 \quad \text{for ice} \tag{2}$$

where $s_o(s'_o)$ is the entropy of water (ice) at 0° C. s_o is related to s'_o and ℓ the latent heat by the following relation

$$s_o - s'_o = \frac{\ell}{273} = \frac{80 \times 4.19 \times 10^3}{273} = 1227. \tag{3}$$

Using (3) and (1) we find

$$s - s'_o = 4222\, \ell n \frac{263}{273} + 1453. \tag{4}$$

From (2) and (4) we find

$$s - s' = 2110\, \ell n \frac{263}{273} + 1528 = 1450 > 0.$$

Since $s > s'$, ice is in the more ordered state.

(V-15) From the lens formula

$$\frac{1}{p_1} + \frac{1}{q_1} = \frac{1}{f_1} \tag{1}$$

with $p_1 = 30$ cm, $f_1 = 20$ cm, we get q_1 equals 60 cm. The size of the image is $5 \times q_1/p_1$ or 10 cm high. The object distance for the second lens is

$$p_2 = 10 - 60 = -50 \text{ cm}. \tag{2}$$

From (1) and (2), we have

$$q_2 = \frac{1}{(1/30 + 1/50)} = \frac{150}{8} \text{ cm} = 18.8 \text{ cm}.$$

The size of the image is

$$10\text{ cm} \times \frac{q_2}{p_2} = 10 \times \frac{18.8}{-50} = -3.75\text{ cm.}$$

Thus, the image is 3.75 cm high and it is inverted.

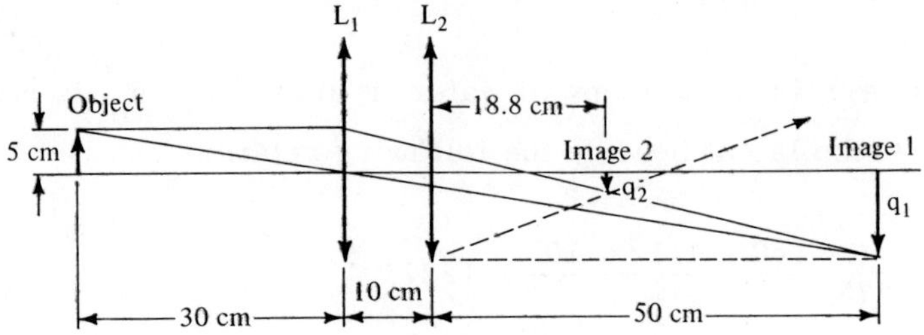

(VI-1) The condition for bright fringes is

$$2nd = (m + \frac{1}{2})\lambda \qquad m = 1, 2, 3, \ldots \tag{1}$$

In going from one fringe to the next the optical thickness of the wedge, nd, should change by $\lambda/2$. Therefore

$$\lambda = 2n(\Delta d) = 2n(\Delta D) \tan\theta \simeq 2n(\Delta D) \times \theta$$

$$= 2 \times 1.4 \times 0.25 \times \frac{20}{60 \times 60 \times 57} \times 10^8$$

~ 7000 Å, where $\Delta D = 0.25$ cm is the distance between fringes.

(VI-2) Let t_1 and t_2 be the instantaneous temperatures of object 1 and object 2 and dW by the amount of work output from the system when dQ_1 and dQ_2 are added to each object, respectively. The maximum attainable efficiency between two heat reservoirs at temperature t_1 and $t_2 (t_1 > t_2)$ is

$$\text{Eff} = \frac{t_1 - t_2}{t_1} = -\frac{dW}{dQ_1} = -\frac{dW}{C_p dt_1}$$

where the minus sign denotes that dQ_1 is actually flowing out of object 1. Therefore

$$-dW = C_P dt_1 \frac{t_1 - t_2}{t_1}. \tag{1}$$

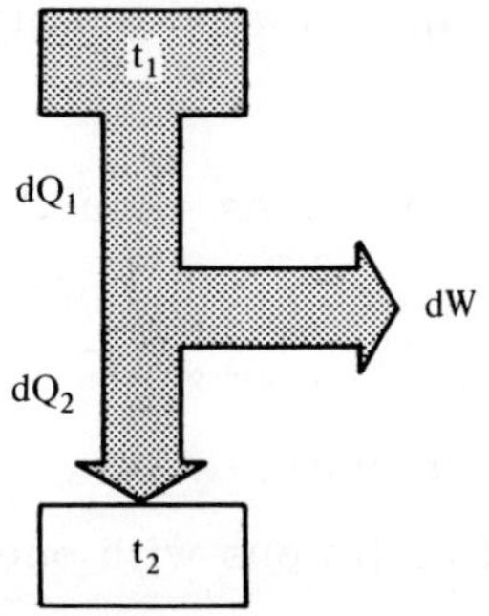

According to the first law of thermodynamics the total energy flowing out of object 1 should equal the sum of the output work from the system and the heat flowing into object 2, i.e.,

$$C_P(T_1 - t_1) = W + C_P(t_2 - T_2)$$

or

$$-C_P t_2 = W + C_P(t_1 - T_1 - T_2). \tag{2}$$

Substituting (2) into (1), we get

$$dW = -\frac{dt_1}{t_1}(W + C_P(2t_1 - T_1 - T_2))$$

or

$$d(Wt_1) = -C_P dt_1(2t_1 - T_1 - T_2). \tag{3}$$

Integrating from T_1 to T_f, we get from (3)

$$WT_f = C_P(T_1^2 - T_f^2 - (T_1 + T_2)(T_1 - T_f)) \tag{4}$$

where T_f is the final temperature of body 1 and body 2 and $W = 0$ when $t_1 = T_1$. Setting $t_2 = t_1 = T_f$ in (2), we can solve for T_f

$$T_f = \frac{1}{2}(T_1 + T_2 - \frac{W}{C_P}). \tag{5}$$

From (5) and (4) it is straightforward to solve for W.

(VI-3)

a. The number of collisions = (number of atoms per unit volume) × (4 times of the cross section of a single atom) × (average velocity of atoms) = $(3 \times 10^{19}) \times 4\pi \times 10^{-16} \times 10^{5} = 4 \times 10^{9}$ per second. (We have used cgs units.)

b. The number of collisions with one wall = (number of atoms per unit volume) × (average velocity of the atoms)/2 = 1.5×10^{24} per second.

(VI-4)

a. The condition determining μ is

$$\sum_i \overline{N}_i = n$$

Using the given expression for $\overline{N}_i$, we get

$$\frac{2}{(2\pi h)^3} \int_0^\infty \frac{1}{e^{\beta(E-\mu)} + 1} d^3p = n, \tag{1}$$

where the factor of 2 comes from the fact that the electrons have spin 1/2. Since $E = p^2/2m$, (1) becomes

$$\frac{2}{(2\pi\hbar)^3} \int_0^\infty \frac{1}{e^{\beta(p^2/2m - \mu)} + 1} 4\pi p^2 dp = n \tag{2}$$

from which one can solve for μ. For $e^{\mu\beta} \ll 1$, we can expand the integrand in power series of $e^{\mu\beta}$ and obtain the relation

$$2\left(\frac{mkT}{2\pi\hbar^2}\right)^{3/2}\left(e^{\mu\beta} - 2^{-3/2} e^{2\mu\beta} + \dots\right) = n \tag{3}$$

b. The thermal wavelength is defined to be

$$\lambda = \sqrt{\frac{2\pi\hbar^2}{mkT}}$$

which is the de Broglie wavelength of a particle with mass m and energy πkT. When $n\lambda^3 \ll 1$, we can neglect the higher-order terms in (3) and find

$$e^{\mu\beta} = \frac{n\lambda^3}{2} = \frac{N}{2v}\lambda^3 \ll 1 \tag{4}$$

Hence the average occupation number $\overline{N}_i$ reduces to

$$\overline{N}_i \propto e^{-\beta(E_i-\mu)} = \frac{N}{2}\frac{\lambda^3}{v} e^{-\beta E_i}.$$

c. At room temperature, we find

$$kT \sim \frac{1}{40}\text{ eV} \quad \text{and} \quad m_e c^2 \sim 0.5 \times 10^6 \text{ eV}, \quad \hbar c = 1973 \text{ eV-Å}.$$

Therefore

$$\lambda \sim \sqrt{\frac{6.3 \times (1973)^2}{0.5 \times 10^6 \times 1/40}} \sim 44 \text{ Å}$$

hence

$$\lambda^3 n \sim (10^{-19})(10^{24}) \sim 10^5.$$

Therefore Maxwell-Boltzmann statistics do not hold for the electron gas.

(VI-5) The plane-polarized wave can be decomposed into two circularly polarized waves of opposite senses but same amplitudes. After passing through the medium, the positive circularly polarized wave is leading the negative circularly polarized wave by a phase Δ:

$$\Delta = (n_+ - n_-)\frac{L}{\lambda} = \left(\frac{K}{\omega(\omega - \Omega)} - \frac{K}{\omega(\omega + \Omega)}\right)\frac{L}{\lambda}$$

$$= \frac{KL}{\omega\lambda}\,\frac{2\Omega}{\omega^2 - \Omega^2}.$$

The plane of polarization is rotated by an angle equal to $\Delta/2$.

(VI-6) The rate of increase of the optical path is

$$(n - 1)\frac{dh}{dt}$$

from which we see

$$\frac{\Delta\nu}{\nu} = \frac{(n - 1)}{c}\frac{dh}{dt}.$$

This problem is similar to the case that the source of light is moving away with velocity

$$v = (n - 1)\frac{dh}{dt}.$$

(VI-7) Let x be the thickness of the plate. The condition of a quarter-wave plate is

$$(n_e - n_o)\frac{x}{\lambda} = \frac{1}{4}$$

or

$$x = \frac{5.8299 \times 10^{-5}}{4 \times (0.01154)} = 1.263 \times 10^{-3}\,\text{cm}.$$

(VI-8)

a. The phase difference of two neighborhing lines after a distance d is

$$\delta = 2\pi\left(\frac{d}{\lambda} - \frac{d}{\lambda'}\right)$$

$$\sim \frac{2\pi d}{\lambda^2}\Delta\lambda$$

where d equals 2 x 30 = 60 cm, $\Delta\lambda$ is the difference between wavelengths, and δ equals 2π for two neighborhing lines. Therefore we can solve for $\Delta\lambda$

$$\Delta\lambda = \frac{\lambda^2}{d} \sim 6 \times 10^{-3}\,\text{Å}.$$

b. The resolving power needed is

$$\frac{\lambda}{\Delta\lambda} = 10^6 .$$

The order of interference, m, is given by

$$m = \frac{2d}{\lambda} = \frac{2 \times 1.5 \times 10^8}{6000} \quad \text{or} \quad 50000.$$

From the given formula, we find the reflectance has to satisfy the condition

$$r^2 + \frac{\pi r}{20} - 1 \geq 0$$

$$(r - 1 + \frac{\pi}{40})(r + 1 + \frac{\pi}{40}) \geq \frac{\pi^2}{(40)^2} \approx 0$$

or

$$r \gtrsim 1 - \frac{\pi}{40} = 0.92$$

in order to resolve the two lines in a.

(VI-9)

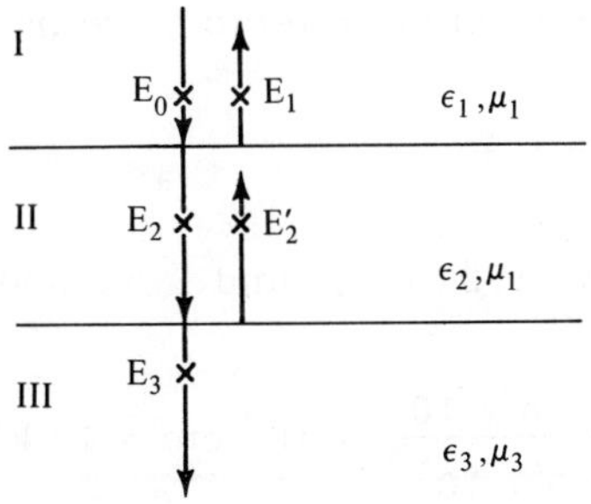

Let E_o, E_1; E_2, E_2'; and E_3 be the electric field in region I, II, and III respectively. For normal incidence, the rays are rotational symmetric. Therefore we can assume the electric fields are all perpendicular to the plane of the paper. The boundary conditions are that the components of the electric and magnetic fields parallel to the surface in one medium are equal to the corresponding parallel components on the other side of the boundary at any instant. Therefore we have the following relations:

$$E_o + E_1 = E_2 + E_2'$$

$$E_2 + E_2' = E_3$$

$$\sqrt{\varepsilon_1}(E_o - E_1) = \sqrt{\varepsilon_2}(E_2 - E_2')$$

$$\sqrt{\varepsilon_2}(E_2 - E_2') = \sqrt{\varepsilon_3}\, E_3$$

from which we can solve for E_1 and E_3 in term of E_o:

$$E_1 = \frac{1 - \sqrt{\varepsilon_3/\varepsilon_1}}{1 + \sqrt{\varepsilon_3/\varepsilon_1}} E_o = \frac{\sqrt{\varepsilon_1} - \sqrt{\varepsilon_3}}{\sqrt{\varepsilon_1} + \sqrt{\varepsilon_3}} E_o$$

$$E_3 = \frac{2E_o}{1 + \sqrt{\varepsilon_3/\varepsilon_1}} = \frac{2\sqrt{\varepsilon_1}\, E_o}{\sqrt{\varepsilon_1} + \sqrt{\varepsilon_3}}$$

which are independent of ε_2. The intensity of the wave transmitted into medium 3 is proportional to E_3^2 while the intensity of the wave reflected back is proportional to E_1^2.

(VI-10) According to Rayleigh criterion, we have

$$\frac{D}{\ell} = 1.22\frac{\lambda}{d} \quad \text{or} \quad \ell = \frac{Dd}{1.22\pi}.$$

Using $d = 0.5$ cm, $D = 1.22$ cm, and $\lambda = 5000$ Å, we find

$$\ell \simeq \frac{1.22 \times 10^2 \times 0.5}{1.22 \times 5 \times 10^{-5}} = \frac{5 \times 10}{5 \times 10^{-5}} = 10^6 \text{ cm} = 10 \text{ kilometers.}$$

(VI-11)

a. 90°.

b. $\varphi = \tan^{-1}\frac{n_1}{n_2}$, where n_1 and n_2 are index of refraction for first and second dielectric.

(VII-1) Red light.

(VII-2) Center of curvature.

(VII-3) Perpendicular to the reflection plane.

(VII-4) If the beam comes from the left, the point of convergence is shifted to the right side of P.

(VII-5) Beyond the focal point and on the same side as the object.

(VII-6) Red.

(VII-7) The inverse of the focal length is

$$(n - 1)\left(\frac{1}{r_1} - \frac{1}{r_2}\right).$$

Since $n_R < n_v$, we have $f_R > f_v$.

(VII-8) Virtual.

(VII-9) Chromatic aberration.

(VII-10) Light cannot pass through the system of a linear polarizer and a linear polarization analyzer when the axes of the two devices are perpendicular to each other.

(VII-11) $\pi/2$.

(VII-12) The ray which passes through the optical center of the lens will not be deviated.

(VII-13) 4000 Å → 7000 Å corresponding to 7.5×10^{14} Hz down to 4.3×10^{14} Hz.

(VIII-14) $\omega_{in} = \omega_{refracted}$; $\lambda_{in} = n\lambda_{refracted}$.

(VII-15) The rays from the center of a sphere are not deviated. Therefore the image must appear at the same plane as the original letter. For small θ, $\theta' = n\theta$ (Snell's law). Since

$\ell' = r\theta' = nr\theta = n\ell$, the size of the image is n times as large as that of the object.

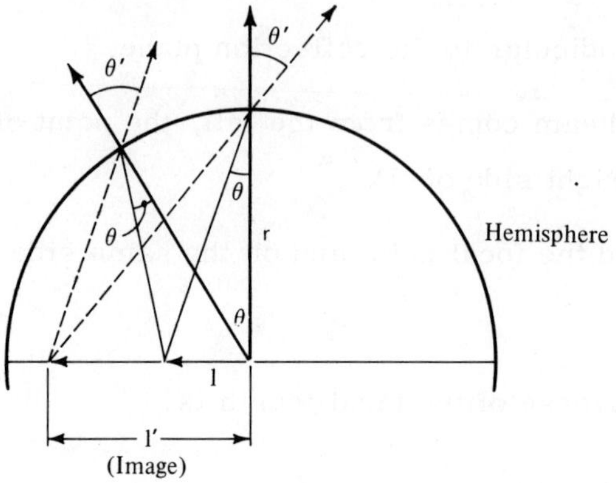

(VII-16)

a. The phase difference of the rays from the two slits at C is

$$\Delta = \frac{2\pi n\delta}{\lambda} - \frac{2\pi\delta}{\lambda} = (n - 1)\frac{2\pi\delta}{\lambda}.$$

The intensity at point C is proportional to $\cos^2 \Delta/2$, or

$$I_c = I_o \cos^2 ((n - 1)\frac{\pi\delta}{\lambda})$$

b. The intensity is minimum for $\delta = (\frac{1}{n - 1})\frac{\lambda}{2}$.

c. The amplitude at point C is

$$A_c = 2Ae^{i\omega t} + Ae^{i(\omega t+\Delta)}$$

$$= Ae^{i\omega t}(2 + e^{i\Delta})$$

from which we find the intensity is

$$I_c \propto |A_c|^2 \propto 4 + 1 + 4 \text{ Re } [e^{i\Delta}]$$

$$\propto 5 + 4 \cos \Delta$$

$$\propto 5 + 4 \cos \frac{2\pi\delta (n - 1)}{\lambda}.$$

(VII-17) This problem is identical in character to Fresnel diffraction by a circular obstacle. The wavelength of the sound is $\lambda = v/f = 0.1$ m. Since the source is far away from the slab, the waves arriving at various points on the plane of the slab are all in phase. For a point on the axis of the slab, the radius of the ℓth Fresnel zone is given by

$$R_\ell^2 + D^2 = (D + \ell \frac{\lambda}{2})^2$$

or

$$R_\ell = (\ell D\lambda + \frac{\ell^2 \lambda^2}{4})^{\frac{1}{2}}$$

where D is the distance between the slab and the listener. The projected area of the ℓth zone is

$$(\pi R_\ell^2 - \pi R_{\ell-1}^2) \cos\theta \approx \pi(\lambda D + \frac{\ell\lambda^2}{2}) \cos\theta$$

where

$$\cos\theta = \frac{D}{\sqrt{D^2 + R_\ell^2}}.$$

The resultant amplitude of the sound wave along the axis is

$$A = \Sigma (-1)^\ell A_\ell$$

where

$$A_\ell = \frac{\pi(\lambda D + \frac{\ell\lambda^2}{2}) \cos\theta}{\sqrt{R_\ell^2 + D^2}} \cos\theta$$

$$= \frac{\pi(\lambda D + \frac{\ell\lambda^2}{2}) D^2}{(R_\ell^2 + D^2)^{3/2}} \tag{1}$$

The components perpendicular to the axis vanish because of rotational symmetry around the axis.

a. For the case that the listener is right behind the center of the slab, we have $D = 0$. Therefore $A_\ell = 0$ for $\ell \neq 0$. Since the first Fresnel zone outside the slab has $\ell = \frac{1}{\lambda/2} = 20$, the resultant amplitude of the sound wave vanishes. Therefore we do not hear the whistle.

b. Now if we go far downstream from the slab, we have $D \gg \lambda$. The first Fresnel zone outside the slab is $\ell_1 = 1/\lambda D$. The resultant amplitude becomes

$$A = \sum_{\ell=\ell_1}^{\infty} (-1)^\ell A_\ell = \sum_{\ell=\ell_1, \ell_1+2, \ldots}^{\infty} (-1)^\ell (A_\ell - A_{\ell+1})$$

$$\approx \frac{1}{2}(-1)^{\ell_1} \sum_{\ell=\ell_1}^{\infty} \frac{dA_\ell}{d\ell} \approx \frac{1}{2}(-1)^{\ell_1} \int_{\ell_1}^{\infty} \frac{dA_\ell}{d\ell}\, d\ell$$

$$= \frac{1}{2}(-1)^{\ell_1+1} (A_{\ell_1} - A_\infty) = \frac{1}{2}(-1)^{\ell_1+1} A_{\ell_1} \tag{2}$$

since $A_\infty = 0$. In the case that the slab is removed, the resultant amplitude is

$$B = \sum_{\ell=0}^{\infty} (-1)^\ell A_\ell \approx \frac{A_o}{2}. \tag{3}$$

The condition that the sound intensity is reduced by a factor of 2 due to the existence of the slab is

$$A^2 = \frac{1}{2} B^2$$

or

$$A_{\ell_1}^2 = \frac{1}{2} A_o^2.$$

Using (1), we get

$$\left\{\frac{\pi(\lambda D + \frac{\ell_1\lambda^2}{2})D^2}{(R_{\ell_1}^2 + D^2)^{3/2}}\right\}^2 = \frac{1}{2}\left(\frac{\pi\lambda D^3}{D^3}\right)^2.$$

Since $R_{\ell_1} = 1$ m, $\ell_1\lambda = \frac{1}{D}$, we find

$$\left\{\frac{D^3 + D/2}{(1 + D^2)^{3/2}}\right\}^2 = \frac{1}{2}$$

or

$$D^6 - D^4 - 2.5D^2 - 1 = 0$$

from which we find $D \approx 1.5$ m.

(VII-18) The size of the spot is approximately the size of the central peak in the diffraction pattern produced by a circular aperture. The first dark ring appears at

$$\sin\alpha = 1.22\frac{\lambda}{d}$$

where

$$\sin\alpha \sim \frac{\text{radius of the spot}}{\text{distance between earth and the moon}} = \frac{R}{D}.$$

With $\lambda = 6000$ Å, $d = 2$ m, and distance $D = 400,000$ km we find radius of the spot

$$R = 1.22\frac{D\lambda}{d} = \frac{1.22 \times 4 \times 10^8 \times 6 \times 10^{-7}}{2} \simeq 1.5 \times 10^2 \text{ m}.$$

(VIII-1) $\overline{V_x} = 0$, since V_s is an odd function.

$$\overline{V_x^2} = \frac{kT}{m} \quad \text{since } \overline{V_x^2} = \overline{V_y^2} = \overline{V_z^2} = \frac{1}{3}\overline{V^2} = \frac{1}{3}[\frac{3kT}{m}]$$

$\overline{V_x^3} = 0$, since V_x^3 is an odd function.

(VII-2) It decreases, since the increase in temperature results in an increase in the viscosity of the air.

(VIII-3)

$$T_f = \frac{T_1 + 5T_2}{6}.$$

The total entropy change of the milk and coffee is

$$\Delta S = 50 \int_{T_1}^{T_f} \frac{c_v\, dT}{T} + 250 \int_{T_2}^{T_f} \frac{c_v\, dT}{T}$$

$$= 50c_v \ln \frac{T_f}{T_1} + 250c_v \ln \frac{T_f}{T_2}.$$

$$= 50 \ln \frac{T_1 + 5T_2}{6T_1} + 250 \ln \frac{T_1 + 5T_2}{6T_2}.$$

because $c_v = 1$ for water.

(VIII-4) The equation of state of an adiabatic process is

$$P_1 V_1^\gamma = P_2 V_2^\gamma = \text{constant}$$

where γ equals 5/3 for monatomic gases. For $V_2 = (1/2)V_1$ the ratio of the pressures is

$$\frac{P_2}{P_1} = \left(\frac{V_1}{V_2}\right)^\gamma = 2^{5/3} \simeq 3.17.$$

(VII-5) From the definition of entropy we have

$$S = \int_0^T \frac{C dT}{T} = \gamma \int_0^T dT = \gamma T.$$

(VIII-6)

$$\overline{(\nu - \nu_o)^2} = \frac{\overline{v_z^2}}{c^2}$$

$$= \frac{kT}{mc^2} .$$

(VIII-7) The change in entropy is $\Delta S = L/T$. According to the first law of thermodynamics we obtain

$$\Delta U = T\Delta S - P\Delta V = L - P(\frac{1}{\rho_2} - \frac{1}{\rho_1}).$$

(VIII-8) Since $N_2 \propto Ne^{-E/kT}$, $N_1 \propto N$, and $N_1 + N_2 = N$, we have

$$C = \frac{dE_{total}}{dT} = E\frac{dN_2}{dT} = E\frac{d}{dT}(\frac{Ne^{-E/kT}}{1 + e^{-E/kT}})$$

$$= NE^2 \frac{\frac{1}{kT^2} e^{E/kT}}{(e^{E/kT} + 1)^2} \longrightarrow \frac{NE^2}{4kT^2} \quad \text{as} \quad T \to \infty$$

and similarly

$$C \longrightarrow NE^2 \frac{1}{kT^2} e^{-E/kT} \quad \text{as} \quad T \to 0.$$

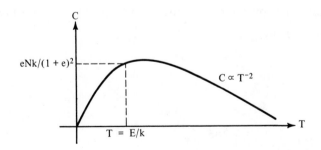

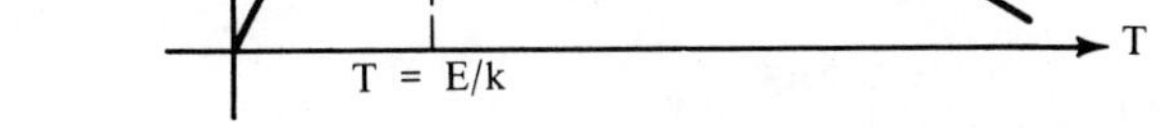

(VIII-9)

a. (3/2)R; for $T \ll \Theta_{rot}$

b. (5/2)R; for $T \gg \Theta_{rot}$

where Θ_{rot} is the characteristic temperature for rotation of the given molecules.

(VIII-10) The area under the curve is multiplied by the factor

$(1/2)^4 = (1/16)$, and the value of λ corresponds to the peak increases by a factor of 2.

(VIII-11)

a. The heat capacity of the iron is

$$C_{iron} = 200 \times 0.60 \times 10^3 = 1.20 \times 10^5 \text{ joules/deg.}$$

The highest efficiency obtainable between two reservoirs of temperature T and T_1 ($T_1 = 285°K = 12°C$) is

$$\text{Eff} = \frac{T - 285}{T}.$$

The total work available is

$$W = C_{iron} \int_{285}^{1773} \frac{T - 285}{T} dT = C_{iron} \int_{285}^{1773} (1 - 285 \frac{dT}{T})$$

$$= (1488 - 285 \ln \frac{1773}{285}) C_{iron} = (1488 - 521) C_{iron}$$

$$= 967 \times 1.20 \times 10^5 = 1.16 \times 10^8 \text{ joules.}$$

b. The entropy change of the iron is related to the heat transferred ΔQ by

$$\Delta S = \frac{\Delta Q}{T}.$$

Therefore

$$S_f - S_i = \int \frac{\Delta Q}{T} = C_{iron} \int_{1773}^{285} \frac{dT}{T}$$

$$= C_{iron} \ln \frac{1773}{285} = - (\ln 6.22) C_{iron}$$

$$= -1.83 \ C_{iron}.$$

The entropy change of the seawater is

$$S'_f - S'_i = \frac{1488 C_{iron}}{285} = 5.22 \ C_{iron}$$

The total entropy change is the sum of the two.

$$(\Delta S)_{Tot} = 3.39C_{iron} = 4.07 \times 10^5 \text{ joule/K.}$$

(VIII-12)

a. Let δ be the phase of the reflected light with respect to the phase of the incident light of amplitude A. The reflected and the transmitted rays are related to A by

$$T = \frac{A}{\sqrt{2}}; \quad R = \frac{A}{\sqrt{2}} e^{i\delta}. \tag{1}$$

According to the reversibility of light rays, if we reverse the directions of the reflected ray and the transmitted ray, we should get the original ray. As shown in the figure we have the relations

$$RR^* + TT^* = A^2$$

and

$$TR^* + RT^* = 0 \tag{2}$$

where

$$T^* = T \text{ and } R^* = \frac{A}{\sqrt{2}} e^{-i\delta}.$$

Using (1) we get for (2)

$$e^{i\delta} + e^{-i\delta} \equiv 2\cos\delta = 0$$

which suggests that $\delta = \pm \pi/2$.

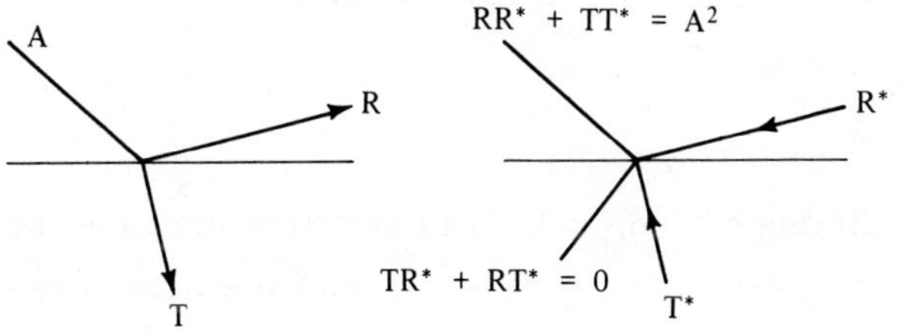

b. The ray BEG is reflected once from the halfsilvered mirror E at E; the phase difference of rays BEG and BEF at E is then $\pi/2$. Similarly, the phase difference at E between ray BCDEF and BCDEG is $\pi/2$. Therefore if rays BEG and BCDEG are out of phase at E, the rays BEF and BCDEF are in phase. In mathematical symbols we find

$$\delta(\text{BEG}) - \delta(\text{BEF}) = m(2\pi) \pm \frac{\pi}{2}$$

$$\delta(\text{BCDEF}) - \delta(\text{BCDEG}) = n(2\pi) \pm \frac{\pi}{2}$$

$$\delta(\text{BCDEG}) - \delta(\text{BEG}) = \ell(2\pi) + \pi \quad \text{(out of phase).}$$

Since we have to choose the same signs (i.e., $\pm\frac{\pi}{2} \pm \frac{\pi}{2} + \pi = {2\pi \atop 0}$) for the same types of mirrors, it follows that

$$\delta(\text{BCDEF}) - \delta(\text{BEF}) = (\ell + m + n + 1)2\pi \quad \text{(in phase).}$$

(VIII-13)

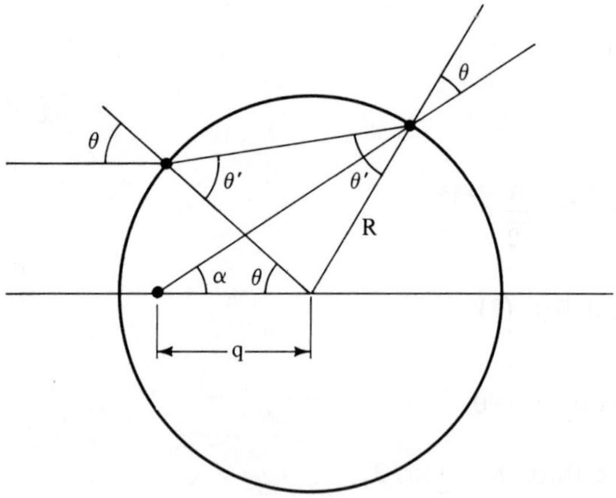

a. According to Snell's law we have

$$\frac{\sin\theta'}{\sin\theta} = n = \frac{4}{3}. \tag{1}$$

The image distance, $-q$, which is negative because the image is on the same side as the object, is related to the radius according to the sine law for a triangle by

$$\frac{q}{\sin\theta} = \frac{R}{\sin\alpha} \tag{2}$$

where $\alpha = \pi - (\pi - 2\theta' + \theta) - \theta = 2(\theta' - \theta)$. Therefore we have

$$q = R\frac{\sin\theta}{2\sin(\theta' - \theta)\cos(\theta' - \theta)}. \tag{3}$$

For rays close to the axis, both θ and θ' are very small. Therefore $\sin\theta \sim \theta$, $\sin\theta' \sim \theta'$, $\sin(\theta' - \theta) \sim (\theta' - \theta)$, and $\cos(\theta' - \theta) \sim 1$. Thus (3) becomes

$$q = \frac{R\theta}{2(\theta' - \theta)} \tag{3)'$$

furthermore, (1) becomes

$$\theta' = (4/3)\theta. \tag{4}$$

Substituting (4) into (3) we obtain

$$q = \frac{3}{2}R. \tag{5}$$

The image distance is just $-q$. The image is on the same side as the object. So it is virtual and erect. The distance, q, by definition is the focal length of the bubble since the object is at infinity.

Alternatively, one can also use the thick lens formulas which were derived for paraxial rays from a point source refracted at a spherical surface of radius R_1 or R_2. We find

$$\frac{n'}{f'_1} = \frac{n' - n}{R_1} \tag{1}$$

and

$$\frac{n}{f'_2} = \frac{n - n'}{R_2} \tag{2}$$

where f'_1 and f'_2 are the focal lengths of the first surface (the left hemisphere) and the second surface (the right hemisphere) of the lens, respectively. For

$$n = \frac{4}{3}, \quad n' = 1, \quad R_1 = R, \text{ and } R_2 = -R$$

we get

$$f'_1 = -3R \text{ and } f'_2 = -4R. \tag{3}$$

The focal length of the lens is given by

$$\frac{n}{f} = \frac{1}{f'_1} + \frac{n}{f'_2} - \frac{2Rn}{f'_1 f'_2}$$

or

$$f = \frac{f'_1 f'_2}{\frac{f'_2}{n} + f'_1 - 2R}.$$

Using (3) and (4) we find

$$f = \frac{-3R}{2}. \tag{5}$$

Finally we can solve for q using the Gaussian lens formula

$$\frac{1}{p} + \frac{1}{q} = \frac{1}{f} \tag{6}$$

where $p = \infty$ is given. Therefore

$$q = f = -\frac{3}{2}R$$

b. For $p = R$ we have

$$\frac{1}{q} = \frac{1}{f} - \frac{1}{R} = -\frac{1}{R}(\frac{2}{3} + 1) = -\frac{5}{3R}$$

therefore

$$q = -\frac{3}{5}R.$$

ATOMIC PHYSICS AND QUANTUM MECHANICS

(I-1) (20 points)

(a) In Rutherford scattering of α particles in a thin gold foil, one neglects the effect of the atomic electrons on the α particle. Why?

(b) In Compton scattering, one neglects the effect of the nucleus on the X-ray. Why?

(I-2) (20 points) Calculate the magnetic field at the proton which arises from an electron in the 2p state of an H atom.

(I-3) (20 points) A mercury lamp emits 10^{18} photons/sec in the 2537° A resonance line. The mercury vapor has low density and is assumed to be at thermal equilibrium at T = 300° K. Calculate the Doppler width of the line. Estimate the natural width. How much power is radiated in the line?

(I-4) (20 points) Potassium is an alkali metal with Z = 19. What is the electron configuration of the ground state of potassium? What are the L, S, and J quantum numbers of the state? Describe the Zeeman effect of the ground state quantitatively.

(I-5) (20 points) Consider an infinite square well. The wave function of a particle trapped in an infinite square well potential of width 2a (see figure) is found to be:

$$\psi = C \cos \frac{\pi x}{2a} + \sin \frac{3\pi x}{a} + \frac{1}{4} \cos \frac{3\pi x}{2a} \quad \text{inside the well}$$

$\psi = 0$ outside the well

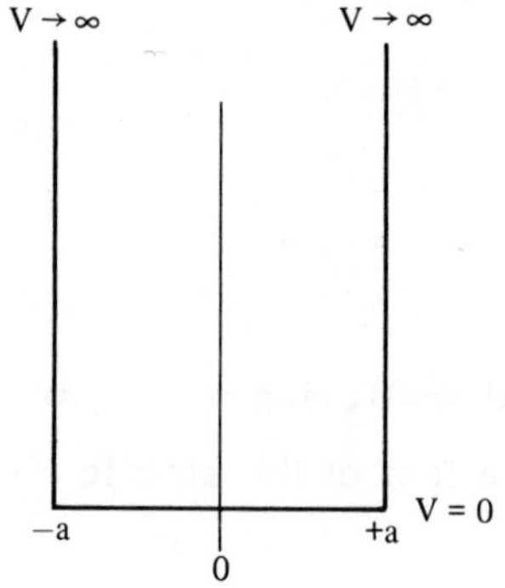

(a) Calculate the coefficient C.

(b) If a measurement of the total energy is made, what are the possible results of such a measurement, and what is the probability to measure each of them?

(II-1) (15 points) Give numerical values and appropriate units for:

(a) Neutron mass.

(b) Planck's constant.

(c) Fine structure constant.

(d) Electron Compton wavelength.

(e) Classical radius of electron.

(f) Lifetime of the 2p state of hydrogen.

(g) Spin magnetic moment of proton.

(h) Lifetime of free neutron.

(i) Velocity of electron in first Bohr orbit.

Express (c), (d), and (e) in terms of the fundamental constants e, $\hbar$, m_e, and c.

(II-2) (25 points)

(a) The time-independent Schröedinger equation for a particle moving in one dimension in a potential $V(x)$ is:

$$\frac{-\hbar^2}{2m}\frac{d^2\psi(x)}{dx^2} + V(x)\psi(x) = E\psi(x)$$

Suppose $V(x) = V(-x)$ and $\psi(x)$ is nondegenerate. Prove that $\psi(x)$ has definite parity, i.e., that:

$\psi(x) = +\psi(-x)$ (Even parity)

or

$\psi(x) = -\psi(-x)$ (Odd parity).

(b) Consider the potential well shown in the figure:

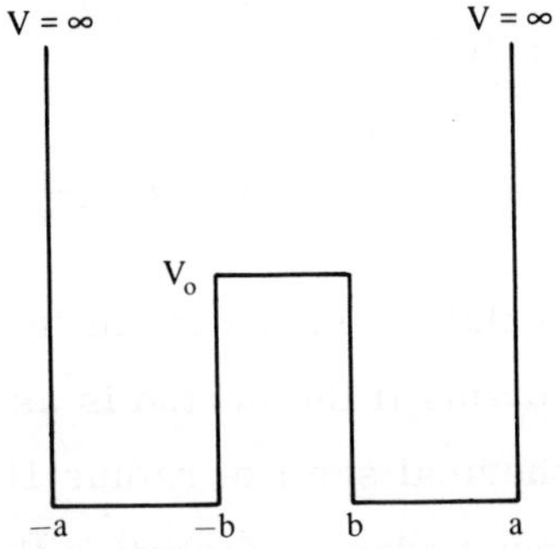

Sketch the approximate character of the two lowest energy solutions to the time-independent Schröedinger equation for this potential. Label the two solutions ψ_1 and ψ_2, and call their respective energies E_1 and E_2.

(c) A particular solution of the time-dependent Schröedinger equation for the potential shown, can be constructed by superposing:

$$\psi_1 e^{-i(E_1/\hbar)t} \quad \text{and} \quad \psi_2 e^{-i(E_2/\hbar)t}$$

Construct a wave packet ψ which at time $t = 0$ is (almost) entirely in the left-hand well. Describe in detail the motion of this wave packet as a function of time.

(II-3) (20 points) The positron has the same mass as the electron, but opposite charge and spin magnetic moment. If you replace the proton in a hydrogen atom by a positron, you get a positronium atom.

(a) The ground state of the atom consists of two hyperfine components, $^1S_0$ and $^3S_1$, split by a very small energy difference. Which level lies lower?

(b) What is the binding energy of the ground state?

(c) A positronium at rest in $^1S_0$ state decays to 2 γ-rays. What is their energy and relative direction?

(II-4) (15 points)

(a) What is the value of the matrix element

$\langle \ell', m' | [L_+, L_-] | \ell, m \rangle$?

(b) Prove that $e^{i\sigma_y \theta/2} = \cos\theta/2 + i\sigma_y \sin\theta/2$.

(II-5) (25 points) Calculate the shift in energy of the 1S state of hydrogen which one obtains if the proton is assumed to be a uniformly charged spherical shell of radius 10^{-13} cm rather than a point charge. Use first-order perturbation theory.

(III-1) (5 points) If the electric charge of the proton were doubled, what would you expect to be the heaviest stable nucleus?

(III-2) (5 points) What order of magnitude of magnetic field strength would be needed to see the Paschen-Back effect rather than Zeeman effect for a typical nucleus? (The effect has never been seen.)

(III-3) (5 points) What minimum kinetic energy must a proton have to produce an antineutron upon colliding with a heavy nucleus?

(III-4) (5 points) Negatively charged muons travelling through matter are very rapidly trapped into Bohr-type orbits about atomic nuclei and are then gradually captured by protons in the nucleus much like inverse beta decay (K capture). The rate at which this capture takes place in various materials follows rather closely a law of the form

muon capture rate = const. $\times Z^p$

Give argument as to why the exponent p should have the value 4.

(III-5) (5 points) Up to what value of the bombarding energy is neutron-proton scattering essentially isotropic?

(III-6) (15 points) What are the energy levels of a particle of mass m moving in one-dimensional potential

$$V(x) = \begin{cases} +\infty & x < 0 \\ +\dfrac{m\omega^2 x^2}{2} & x > 0 \end{cases}$$

No lengthy calculations are needed.

(III-7) (5 points) List all the absolute symmetries (conservation laws) you know. List some approximate ones.

(III-8) (5 points) A particular meson can decay into two different sets of final states at rates given by the separate decay times t_1 and t_2. Give the formula for the uncertainty in mass of this meson.

(III-9) (5 points) What are the values of the phase and group velocities of the deBroglie wave, describing a free electron with classical velocity V?

(III-10) (10 points) Describe for each of the following states the domain of all points in three-dimensional configuration space at which the wave function ψ of the hydrogen atom vanishes. The normalization criterion is

(a) the 1s state

(b) the 2s state

(c) the 2p state (specify here which substate you are describing).

(III-11) (10 points) Give the simplest formula for the binding energy of an electron in the K shell of an atom of nuclear charge Z. Will corrections for

(a) relativistic form

(b) screening of the central field by other electrons

(c) the finite size of the nucleus

increase or decrease the value?

(III-12) (10 points) Give the functional form (with sign) of the variation with separation of the potential energy between the following particles at large distance:

(a) two neutral atoms

(b) two ionized atoms

(c) one neutral atom and one ion

(d) two neutrons (considering nuclear forces only)

(e) two neutrons (considering electromagnetic forces also)

(III-13) (15 points) Consider a homogeneous quantum mechanical sphere whose center is constrained to remain at the origin, but which is free to rotate. Since points on its surface are indistinguishable, its wave function satisfies the condition that $|\psi(\theta,\varphi)|^2$ is independent of the angles θ and φ. Determine the allowed values of the angular momentum and prove your result.

(IV-1) (15 points) A one-dimensional potential barrier is of the shape

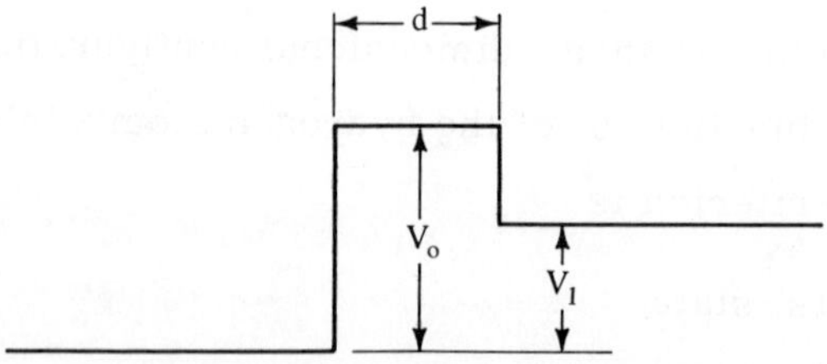

Find the transmission coefficient for particles of mass m coming from the left, with energy $E(V_1 < E < V_0)$.

(IV-2) (5 points) Determine the eigenvalues and normalized eigenvectors of this matrix:

$$\begin{pmatrix} 3 & 2 \\ 2 & 0 \end{pmatrix}$$

(IV-3) (20 points) Assume that Ψ is an eigenfunction of the single-particle Schröedinger equation. Define the vector $\underset{\sim}{C}$ as follows:

$$\frac{d}{dt}\{\int_V \Psi * \Psi \, dv\} = -\int_V \underset{\sim}{\nabla} \cdot \underset{\sim}{C} \, dv$$

(a) What is the physical meaning of $\underset{\sim}{C}$?

(b) Assume that ψ satisfies the Schröedinger equation and calculate an explicit expression for $\underset{\sim}{C}$.

(IV-4) (20 points) Write down the wave function for a hydrogen atom in its ground state. Find an expression for the probability of finding the particle (the electron) between r and $r + dr$, independent of the angular location of electron with respect to origin. Find the value of r at which this probability is maximum.

(IV-5) (20 points) Give the numerical values of the following physical quantities to within a factor of 3.

(a) Distance apart of two protons in a hydrogen molecule.

(b) Wavelength of greatest intensity in black-body radiation at 3° K.

(c) Energy gap between valence and conduction band in pure silicon crystal.

(d) Energy released in fission of a nucleus of uranium 235.

(e) Frequency of red light.

(f) Time required for light wave to traverse diameter of one proton.

(g) Magnetic moment of free electron.

(IV-6) (20 points) A particle of mass m rests in its ground state in a one-dimensional box formed by two very high ("infinite") potential walls separated by a distance a. Suddenly the walls are symmetrically expanded to double the original separation.

(a) What is the probability that the particle now finds itself in the ground state of the expanded system?

(b) Is energy conserved in this process?

(V-1) (20 points) Give a brief discussion of the following topics.

(a) Correspondence principle

(b) Dulong and Petit law

(c) Why only the principal series is observed in the absorption spectra of alkali metals.

(d) Why the Lande g factor is 1 for all singlet states and 2 for all S states.

(e) Two experiments that demonstrate the particle-like behavior of electromagnetic radiation.

(V-2) (20 points) A μ-meson which is 210 times as heavy as an electron is captured by a proton to form a hydrogen-like atom.

(a) What is the energy of the photon that is emitted when the μ-meson falls from the first excited state to the ground state?

(b) What is the radius of the first Bohr orbit?

(c) What is the velocity of the μ-meson in the nth circular Bohr orbit?

n is the principal quantum number.

(V-3) (20 points) In elementary chemistry, students identify the presence of small amounts of sodium by the yellow light (5890 Å) emitted when a sample is placed in a Bunsen flame. This might appear unreasonable in view of the relatively low temperature of the flame (2000° K). Demonstrate quantitatively that this is not the case and explain the result.

(V-4) (20 points) A one-dimensional particle moves in the potential energy

$$V(x) = 0 \qquad x < 0$$
$$= V_0 \qquad x > 0$$

Suppose the particle has energy $E > V_0$ and is incident from the (-x) direction.

(a) Find the wave function everywhere. You do not need to normalize it.

(b) Normalize the wave function so it corresponds to unit incident flux (one particle per second).

(c) Solve the case for $E < V_0$ and discuss the significance of the result.

(V-5) (20 points) The (unnormalized) wave function of a particle of mass m is

$$\psi_k(r) = \frac{e^{-ikr} + be^{ikr}}{r} \qquad \text{(r is radial distance from origin)}$$

(a) What is the energy of the particle?

(b) Is the particle a free particle? If your answer is no, describe the potential as well as you can.

(VI-1) (15 points)

(a) The continuity equation of electromagnetic theory relates the divergence of the current density to the time dependence of the charge density. Use an analogous idea to derive the nonrelativistic probability current density of quantum mechanics.

(b) Compute the probability current density for the unnormalized wave function $\psi = e^{ikx}$.

(c) Can a real wave function have a current?

(VI-2) (20 points) A boy on the top of a ladder of height H is dropping marbles of mass m to the floor. To aim well, the boy is using equipment of the highest possible precision. Estimate, using the uncertainty principle, the typical distance by which he will miss the crack.

(VI-3) (20 points) According to Moseley's law the frequency of the K_α X-ray line is given as a function of Z, the atomic number, by a relation of the form:

$$\sqrt{\nu} = aZ - b$$

(a) Give an approximate formula for a in terms of fundamental constants.

(b) Explain why the frequencies of X-ray lines vary from element to element in a simple way, whereas optical frequencies follow no such simple relation.

(VI-4) (10 points) A particle, of rest mass m_1 and velocity v, collides with a stationary particle of rest mass m_2, and is absorbed by it. Find the rest mass M and velocity V of the resultant particle.

(VI-5) (15 points) Electrons enter a region of uniform magnetic field B through a slit, S_1, and leave, after one semicircular turn, through slit S_2. As they enter S_1, their spins are pointed upwards ($\varphi = 0$) as shown in the diagram. Electrons have a g-factor of $g = 2 + \alpha/\pi$, and the magnetic moment is given by $\vec{\mu} = -\frac{e}{2mc} g\vec{s}$.

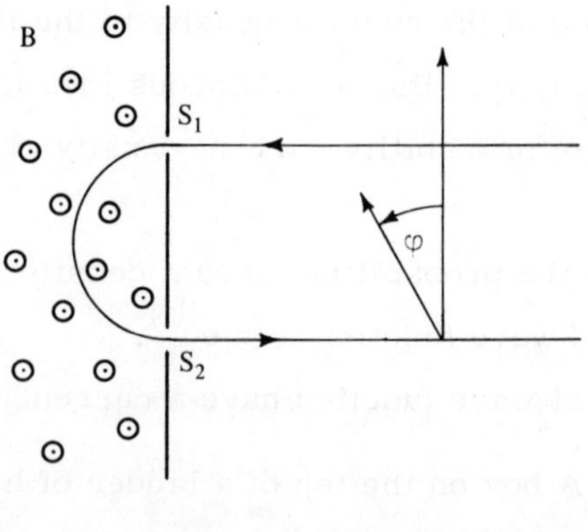

(a) What is their spin precession frequency?

(b) What is their cyclotron frequency?

(c) What is the angle φ, which the electron spins make with the initial spin direction, when they exit through slit S_2?

NOTE: It can be proved that the dynamical equation for the time variation of the quantum expectation value of the electron spin is identical to the classical dynamical equation for the electron spin. Therefore a completely classical approach is valid here and should be used.

(VI-6) (20 points)

(a) Using classical expressions for the kinetic and potential energy of the electron-proton system and the Bohr quantum condition, derive an expression for the energy levels of the hydrogen atom.

(b) Estimate the energy difference between the second and fourth energy levels.

(VII-1) (20 points) An electron with a K.E. of 10 eV at $x = -\infty$ is moving from left to right along the x-axis. The potential energy is $V = 0$ for $x < 0$ and $V = 20$ eV for $x > 0$. Treat the electron as a one-dimensional plane wave.

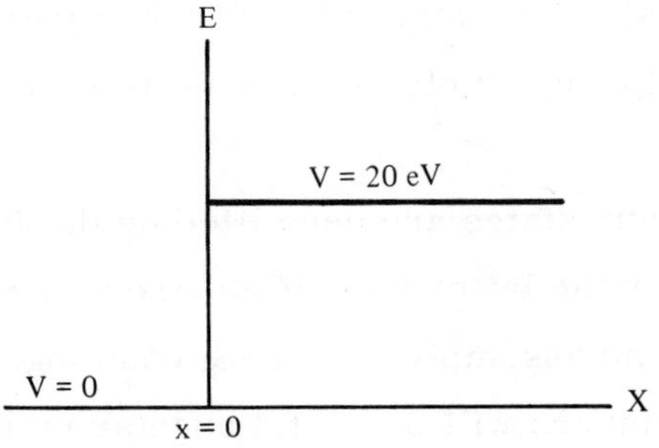

(a) Write the Schröedinger equation for $x < 0$ and $x > 0$.

(b) Sketch the solutions in the two regions.

(c) What is the wavelength for $x < 0$ (in centimeters)?

(d) What are the boundary conditions at $x = 0$?

(e) Make a general statement about the possibility of finding the electron at some positive value of x.

(VII-2) (20 points)

(a) In a one-dimensional harmonic oscillator, with characteristic frequency ν_o, what is the energy and parity of the eigenstate associated with quantum number n? What values may n have?

(b) The wave function of a three-dimensional harmonic oscillator may be written as the product of three one-dimensional harmonic oscillator eigenfunctions of each of the three Cartesian coordinates, with respective quantum numbers $n_x n_y n_z$. Find the energy, parity, and degeneracy of the lowest four distinct groups of energy levels.

(c) The three-dimensional harmonic oscillator can also be solved in spherical coordinates. The same energy eigenvalues, through different eigenfunctions, will be found. Using your knowledge of the parity and degeneracy of various states, deduce which values of angular momentum quantum number, ℓ, are associated with each group of levels enumerated above in part (b).

(VII-3) (20 points) A free atom of carbon has four paired electrons in s-states and two more electrons with p-wave orbital wave functions.

(a) How many states are permitted by the Pauli exclusion principle for the latter pair of electrons in this configuration?

(b) Under the assumption of L-S coupling, the "good" quantum numbers will be: total angular momentum, $\vec{J}$; $L^2 = (\ell_1 + \ell_2)^2$; and $S^2 = (s_1 + s_2)^2$. Give the sets of values of J, L, S allowed for this configuration of two p-wave electrons.

(c) Add up the degeneracies of the terms found in (b), and show that it is the same as the number found in (a).

(VII-4) (20 points) The atomic number of Na is 11.

(a) Write down the electronic configuration for the ground state of the Na atom showing in standard notation the assignment of all electrons to the various one-electron states.

(b) Give the standard spectroscopic notation for the ground state of the Na atom. (Prototype form: $5\,^3F_2$)

(c) The lowest frequency line in the absorption spectrum of Na is a doublet. What are the spectroscopic designations of the pair of energy levels to which the atom is excited as a result of this absorption process?

(d) What is the mechanism responsible for the splitting between this pair of energy levels?

(e) The total angular momentum j of the atom is different for these two levels. Does the level corresponding to the higher value of j lie lower or higher?

(f) The splitting between these two levels is proportional to the average value of r^n, where r is the distance of the valence electron from the nucleus. Derive the power n by a simple argument.

(VII-5) (20 points)

(a) State all the commutation relations among the angular momentum operators L_x, L_y, L_z and L^2.

(b) Let $\psi_{\ell m}$ be an eigenstate of L^2 and L_z with eigenvalues $\hbar^2 \ell(\ell + 1)$ and $\hbar m$, respectively. Show that $\phi = (Lx + iLy)\psi_{\ell m}$ is likewise an eigenstate of L^2 and L_z, and determine the eigenvalues.

(c) Show that if $\ell = 0$, the state $\psi_{\ell m}$ of part (b) is also an eigenstate of L_x and L_y.

(VIII-1) (15 points) State the quantum numbers S, L, J for each of the following:

(a) The ground state of neutral boron (atomic no. 5).

(b) The ground state of singly ionized sodium (atomic no. 11).

(c) The ground state of doubly ionized sodium.

(d) The first excited state of singly ionized sodium.

(e) The ground state of the H_2 molecule.

(VIII-2) (10 points) A particle of mass m is bound in a one-dimensional potential of characteristic width a, and has a ground state energy E = -B. There are two quantities of the dimension of length which occur from a study of Schröedinger equation:

$$d_1 = a \quad \text{and} \quad d_2 = \sqrt{\frac{\hbar^2}{2mB}}$$

Answer qualitatively the question: "Over what distance d is the particle's probability density spread out?" Sketch the wave function in order to illustrate your conclusions.

(VIII-3) (10 points) In the figure, A and B are Stern Gerlach magnet systems which allow only $S_z = +1/2$ states to pass (we have a bunched beam of potassium atoms incident from the left). The bunched beam is subjected to a spatially uniform magnetic field B in the region between A and B for an interval of 1 microsecond. What should be the magnitude and direction of B so that none of the beam can reach the detector D?

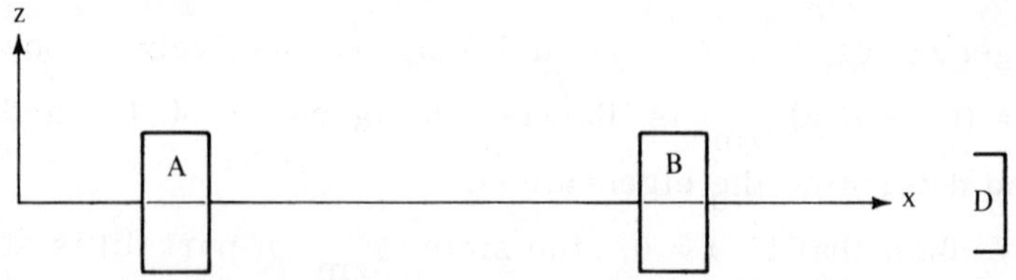

(VIII-4) (10 points) Approximately what is the lifetime τ of the 2p state of hydrogen? According to the uncertainty principle we would say that the energy of this state is uncertain by an amount roughly $\Delta E = \hbar/\tau$. In a famous series of measurements begun twenty years ago, W. E. Lamb determined the energy of this state with an accuracy about one thousand times better than this ΔE. Explain.

(VIII-5) (15 points) Give the formula and numerical value in cm accurate to at least one significant figure for:

(a) The radius of the first Bohr orbit in H.

(b) The radius of the first Bohr orbit in Hg.

(c) Compton wavelength of the electron.

(d) Compton wavelength of the pi-meson.

(e) De Broglie wavelength of a 10 keV neutron.

(f) De Broglie wavelength of a 10 BeV proton.

(g) Compton wavelength of the neutrino. (Specify which kind of neutrino.)

(h) The radius of the heaviest stable nucleus.

(VIII-6) (10 points)

(a) Prove that the expectation value of the momentum operator $\vec{p}$ must vanish in any stationary state.

(b) Under what general conditions can one also prove that the expectation value of the position operator $\vec{r}$ must vanish?

(VIII-7) (10 points) State the selection rules for electric dipole transitions in light atoms. Which of these rules remain valid in heavy atoms? in nuclei?

(VIII-8) (10 points) Atomic hyperfine structure is smaller than atomic fine structure by approximately the ratio m/MZ, where m is the electron mass, M the proton mass, and Z the atomic number. Explain this result.

(VIII-9) (10 points) Consider the function f(x) defined by the power series

$$f(x) = \sum_{n=0}^{\infty} \frac{x^n}{(n!)^p}$$

where p is some positive real number. How fast does this function increase as x goes to + infinity? Could this be an acceptable solution to some Schröedinger equation when we have the normalization requirement

$$\int_0^{\infty} |\psi(x)|^2 dx = 1 ?$$

Explain.

(IX-1) (5 points) A cyclotron accelerates deuterons to 16 MeV energy. The deuterium is replaced by helium. What energy alpha particles are obtained?

(IX-2) (5 points) What is the density in gm/cm^3 for nuclear matter?

(IX-3) (5 points) Alpha particles and protons of the same kinetic energy are passed through a gold foil. What is the ratio of their coulomb scattering (nonrelativistic)?

(IX-4) (5 points) How much energy (in joules) would be released if a meteor of 1 kg of antimatter were to hit the earth?

(IX-5) (5 points) What is the ground state energy of an atom consisting of an electron and a positron bound to each other by their coulomb attraction?

(IX-6) (5 points) How many spectral lines appear in the Zeeman splitting of the $^2D_{3/2} \rightarrow {}^2P_{1/2}$ transition of sodium?

(IX-7) (5 points) A particle of mass m is confined by an appropriate potential to a spherical region of radius R. Give a rough estimate of the particle's kinetic energy in its ground state. (Consider only nonrelativistic quantum mechanics.)

(IX-8) (5 points) Calculate the short wavelength limit of a 25 keV X-ray tube.

(IX-9) (5 points) A large number of identical Fermi particles are confined in a box of volume V, occupying the lowest accessible levels. By what factor does the maximum momentum of the particles change if we double the volume of the box, leaving the number of particles unchanged?

(IX-10) (5 points) State the magnetic moment of the proton to within a factor of 2 in accuracy.

(IX-11) (5 points) What is the ground state (in spectroscopic notation) of the helium atom?

(IX-12) (5 points) A particle moving under the influence of a potential $V = 1/2kr^2$ has a wave function $\Psi(r, t)$. If we change the wave function to $\Psi(\alpha r, t)$, by what factors are the mean kinetic and potential energies changed?

(IX-13) (10 points) Using elementary Bohr theory,

(a) Calculate the magnetic field at the center of a hydrogen atom where the electron is in its ground state.

(b) Estimate the hyperfine splitting of this state.

(c) Calculate the frequency of radio emission that would be observed from hydrogen gas such as distributed throughout the galaxy (neglect the magnetic moment of the electron).

(IX-14) (15 points) Consider an atomic system consisting of two protons and one electron. Suppose first that the protons are held stationary at a distance R from each other, and that we study the ground state energy of the electron as a function of R, $E_o(R)$.

(a) Give expressions for $E_o(\infty)$ and $E_o(0)$.

(b) Suppose now that we consider the two protons to have an

"effective potential energy" $V(R)$ equal to the sum of $E_o(R)$ and their Coulomb repulsion potential energy. In terms of $E_o(R)$, what is the equilibrium position of the protons?

(c) If we assume that $E_o(R)$ changes smoothly and monotonically from $E_o(\infty)$ to $E_o(0)$, make a rough sketch of $V(R)$ as a function of R.

(d) So far we have neglected any kinetic energies of the protons. Describe qualitatively the nature of the low-lying energy states corresponding to the electron in the ground state and the two protons in low excited states.

(IX-15) (15 points) A parallel beam of electrons is accelerated through 37 volts and is incident normally on a screen containing a slit 1.0 Å wide (clearly an impossible situation). A detector of small dimensions (~ 1 Å) scans perpendicular to both the slit and the beam direction 10 cm from the screen:

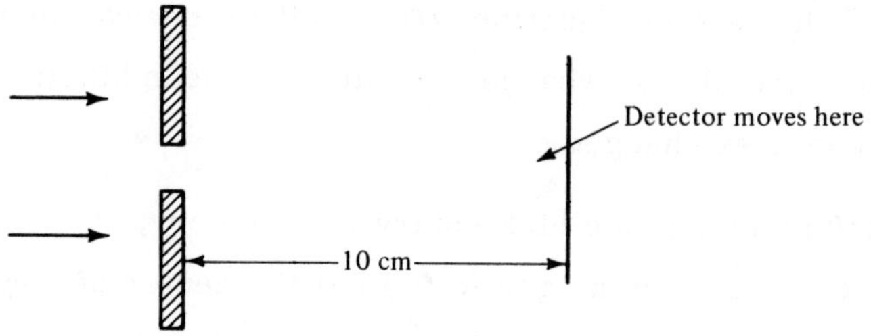

(a) Approximately how wide is the intensity pattern recorded by the detector?

(b) If a second slit similar to the first one is placed in the screen parallel to the first slit and 10 Å away from it, sketch the intensity pattern observed by the detector.

(c) Suppose that the intensity of the electron beam is reduced until only one electron is in the region between the slits and detector at a given instant. How does the pattern change?

(d) What is the observed pattern if a second (transparent)

detector is placed across one slit to obtain knowledge of which slit a particular electron passed through?

(X-1) (20 points)

(a) A muon is a particle of mass 206 m_e and charge equal to that of an electron. If a negative muon is captured by an atom of phosphorous ($Z = 15$) and cascades down the various energy levels, what is the energy of a photon emitted by a transition of the muon between the $n = 3$ level and $n = 2$ level?

(b) Experimentally a precise determination of the energy of the photon emitted as described in (a) can be used to obtain an accurate value of the muon mass. It turns out that the photon energy lies near the middle of the K absorption edge in lead ($Z = 82$). Make a rough calculation of the energy at which the lead K absorption edge occurs to show that the above statement is reasonable.

(c) Explain why, from an experimental point of view, this is a fortunate circumstance and indicate how you would use this fact to obtain a precise measure of the photon energy.

(X-2) (25 points)

(a) Starting with the expression for the energy radiated per second by an accelerating charge,

$$\frac{dw}{dt} = \frac{2}{3}\frac{e^2 a^2}{c^3}\frac{\text{ergs}}{\text{sec}}$$

derive the expression for the Thomson cross section.

(b) Describe how Thomson scattering is related to Compton scattering.

(c) Suppose 1/2 MeV γ-rays are scattered from hydrogen atoms through an angle of 90 degrees.

(1) What is the energy of the photons scattered from the electron? From the proton?

(2) Make an order-of-magnitude-estimate of the ratio of the scattering rates $\gamma + e^- / \gamma + p$.

(X-3) (20 points) Given the classical formula (see problem (X-2)) for the power radiated by an accelerating charge, and using the correspondence principle, find the mean life of a highly excited quantum state of a simple harmonic oscillator in terms of the quantum number n of the state, the classical angular frequency ω, and the mass m and charge e of the oscillating particle.

(X-4) (20 points) A simple model of a nucleus of N neutrons and Z protons considers the nucleons bound in an infinitely deep (square) potential well.

(a) Calculate an expression for the density of energy levels (i.e., the number of levels per unit energy interval) in this potential.

(b) When the nucleus is in its lowest energy state, what is the maximum kinetic energy of a single nucleon?

(c) Show that if the nuclear density is constant this energy is independent of the number of nucleons.

(d) How can the model be modified to take account of the electric forces between protons?

(X-5) (15 points) A paramagnetic salt of $Ti^{3+}[3d^1, {}^2D_{3/2}]$ is placed in a magnetic field of 10,000 gauss and in a bath of liquid helium at 1°K. Approximately what fraction of the Ti ions will have their spins aligned parallel to the field.

(XI-1) (20 points) A javelin is "perfectly" balanced vertically with its point on a stationary horizontal marble slab. Estimate the time it takes to fall over from uncertainty principle considerations.

(XI-2) (20 points) In some magnetic materials there exist spin waves of frequency $\omega = Dk^2$ where D is a constant and k is the wave number (magnitude of the propagation vector) of a given spin wave.

The energy levels are quantized, and given by $E = n\hbar\omega$.

(a) Find the phase velocity v and the group velocity u for these waves as a function of ω.

(b) Derive the temperature dependence of the total energy density U associated with these waves at thermal equilibrium.

HINTS: These waves are bosons and thus have a thermal energy distribution given by Planck's law as does black-body radiation. The number of waves $N(k)$ having wave numbers between k and $k + dk$ can be obtained as in the theory of the free electron gas (except that the spin waves have no factor of 2 for polarization as do the photons and electrons).

(XI-3) (20 points) Give a brief discussion of the following topics in quantum mechanics:

(a) Physical interpretation of the wave function $\Psi(x)$.

(b) Rules for vector addition of angular momenta.

(c) Selection rules for changes of j, ℓ, and m for allowed electromagnetic electric dipole transitions.

(d) Required relationship between any two dynamical quantities G and F that they be simultaneously measurable to arbitrary accuracy.

(XI-4) (5 points) Estimate the kinetic energy of a nucleon in a carbon nucleus. The diameter of the nucleus is about 2×10^{-15} meter.

(XI-5) (5 points) The sun radiates energy at the rate of 4×10^{26} watts including its neutrino emission. Assume the power is generated entirely by the proton-proton cycle, which results in

$$4p \rightarrow He^4 + 26 \text{ MeV}.$$

How many helium atoms are being formed per second in the sun's interior?

(XI-6) (5 points) What is the approximate (i.e., within $\pm$ 30%) energy of K X-rays for copper?

(XI-7) (5 points) What minimum incident energy must an electron have in order to produce an electron-positron pair upon striking a second electron which is at rest?

$e^- + e^- \rightarrow e^- + e^- + e^- + e^+$

(XI-8) (5 points) An atom has atomic number $Z = 26$ and frequently shows valence +2. Give the configuration state of its electrons. [Example: The configuration state of oxygen is: $(1s^2\ 2s^2\ 2p^4)\,^3P_2$.]

(XI-9) (5 points) Estimate the Zeeman splitting $\Delta\nu$ of hydrogen spectral lines in a magnetic field of 10,000 gauss.

(XI-10) (10 points) A particle moves in a symmetrical one-dimensional potential, as illustrated:

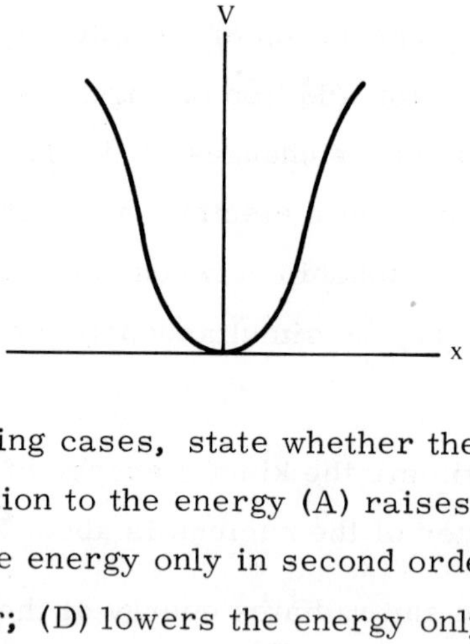

In each of the following cases, state whether the first-order perturbation correction to the energy (A) raises the energy in first order; (B) raises the energy only in second order; (C) lowers the energy in first order; (D) lowers the energy only in second order; (E) doesn't change the energy.

(a) Perturbing potential: 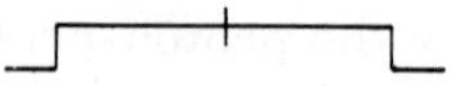 System is in ground state ________.

(b) Perturbing potential: 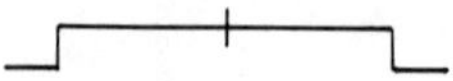 System is in first excited state ________.

(c) Perturbing potential: 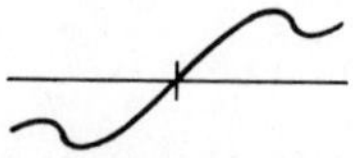 System is in ground state ________.

(XII-1) (20 points) Consider a state of a two-particle system represented by the wave function

$$\Psi = e^{iP(x_1+x_2)/2\hbar}\, e^{-(Mk/2)^{1/2}(x_1-x_2)^2/2\hbar}$$

where x_1, x_2 are the positions of the two equal mass (M) particles moving in one dimension and interacting with a harmonic oscillator force $F = -k(x_1 - x_2)$.

(a) What is the expectation value of total energy of relative motion?

(b) What is the mean absolute value of relative momentum?

(c) If a measurement of relative momentum p were made, with what probability would one obtain

$$p < \sqrt{\hbar\sqrt{2Mk}}.$$

(XII-2) (10 points) The wave function

$$\psi(r) \underset{r\to\infty}{\sim} e^{ikz} + f(\theta)\frac{e^{ikr}}{r}$$

is used to describe a quantum-mechanical scattering process.

(a) Use ψ to find expressions for the incident and scattered probability current.

(b) Find the relationship between $f(\theta)$ and the differential cross section $\sigma(\theta)$.

(XII-3) (10 points) Below is a plot of data (say, the excitation cross section of an atom vs bombarding electron energy). A quick run has been made to obtain the general form.

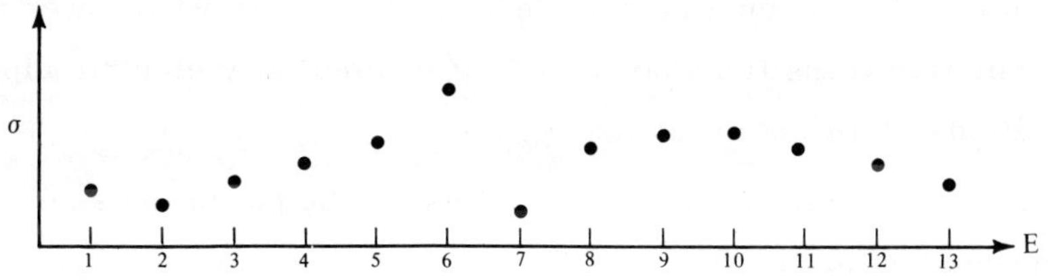

You have time for five more points. Approximately where would you take them and why?

(XII-4) (5 points) What is the probability that a long-lived radioactive source gives 9700 counts or less during a one-second interval if its average counting rate with the same detector is 10^4/sec?

(XII-5) (5 points) What is the probability that a radioactive atom having a mean life of ten days decays during the fifth day?

(XII-6) (5 points) The half-life of the neutron is about 12 minutes. How much energy (in MeV) must a neutron have to give it a 50% probability of surviving a trip to the earth from a star 10 light years away?

(XII-7) (3 points) A K meson decays into two spin zero neutral pions. What are the possible values of the K meson spin?

(XII-8) (3 points) Draw the configuration of least energy for three positive charges constrained to remain on the surface of a sphere. Similarly, for four charges.

(XII-9) (3 points) Estimate the height of the coulomb barrier at the nuclear surface for α-decay of U^{238}.

(XII-10) (3 points) What would be the radius of the 1s orbit of a neutron-electron system bound by the gravitational force only? (The universal gravitational constant $G = 6.7 \times 10^{-8}$ dyne-cm^2/gram^2.)

orbit radius = __________ cm.

(XII-11) (3 points) How does the radius of the K shell of an atom depend upon the atomic number Z?

(XII-12) (3 points) Consider a hydrogen atom whose electron is in the state with quantum numbers $n = 3$, $\ell = 2$. To what lower states are radiative transitions possible? (Consider only electric dipole transitions.)

(XII-13) (3 points) How many electrons can be put in the shell corresponding to $n = 5$?

(XII-14) (3 points) Consider an atomic system consisting of an electron and positron bound to each other in a state of orbital angular momentum $\ell = 1$. What is the magnetic moment arising from the orbital motion?

(XII-15) (3 points) State the eigenvalues of each of the four spin operators for an electron:

s_x, s_y, s_z, and $s^2 = s_x^2 + s_y^2 + s_z^2$.

(XII-16) (3 points) Imagine that an electron is fixed in position so that it has only a spin degree of freedom. Express its energy levels in a uniform magnetic field H in terms of H and the fundamental constants e, $\hbar$, m, c.

(XII-17) (3 points) A particle at rest, whose mass is m, decays into two photons (a π^o mean, for example). The formula for the momentum p of each photon is

p = ________.

(XII-18) (3 points) A photon strikes an electron in a hydrogen atom at rest liberating the electron. Under what condition is the assumption that the electron was initially unbound and at rest a good approximation in analyzing this event?

(XII-19) (3 points) Consider a hydrogen atom in quantum state characterized by principal quantum number n = 2. In the semiclassical picture, which orbit is the more eccentric, $\ell = 0$ or $\ell = 1$?

(XII-20) (3 points) The binding energy of the electron in its ground state in the He^+ ion is, in electron volts: ________

(XII-21) (3 points) Name two important mechanisms by which a beam of 500 keV photons is attenuated while passing through matter.

SOLUTION SET OF

ATOMIC PHYSICS AND QUANTUM MECHANICS

(I-1)

a. The size of the nucleus is small compared with the de Broglie Compton wavelength of low energy α particles. Therefore the scattering of low energy α's by the nucleus is coherent. The cross section for coherent scattering of an α particle from Z-protons in the nucleus is proportional to Z^2. The scattering of an α due to the electrons is incoherent because the radius of an electron's orbit is much larger than the de Broglie wavelength of the particle. The cross section for α-electron scattering is proportional to Z. Therefore the scattering of α-particles by a nucleus dominates over that due to electrons. Furthermore the energy loss due to α-nucleus scattering is much larger than the ionization energy loss to the electrons, since the energy transferred to an electron per α-electron scattering is less than $2m_e v_\alpha^2$. Finally the scattering angle of the α-particle is much larger when it is scattered from a nucleus rather than from an electron. These are the main reasons why one can neglect the effect of the atomic electrons.

b. The cross section of Compton scattering is proportional to the inverse of the square of the mass of the scattering particle. Therefore we have

$$\frac{\text{cross section of photon-electron scattering}}{\text{cross section of photon-nucleus scattering}} = \left(\frac{m_N}{m_e}\right)^2 \gg 1$$

from which we see the effect of the nucleus on the X-ray is negligible.

(I-2) The magnetic field at the center of a loop with current i is

$$B = \frac{\mu_0 i}{2a} \tag{1}$$

where a is the radius of the loop. For an electron in the 2P state of a H atom we have

$$a = n^2 a_o, \qquad v = \alpha \frac{c}{n} \text{ with } n = 2$$

and

$$i = \frac{ev}{2\pi a} = \frac{e\alpha c}{2\pi n^3 a_o} = \frac{1.6 \times 10^{-19} \times \frac{1}{137} \times 3 \times 10^{8}}{2 \times 3.14 \times 4 \times 10^{-10}}$$

$$= 1.4 \times 10^{-4} \text{ amp.}$$

Therefore we find

$$B = \frac{4\pi \times 10^{-7} \times 1.4 \times 10^{-4}}{2 \times (2 \times 10^{-10})} = 0.43 \frac{\text{webers}}{m^2}.$$

(I-3) From classical wave theory we know that the Doppler width is

$$\frac{\Delta\omega_D}{\omega} \sim \frac{\sqrt{\langle V^2 \rangle}}{c} \tag{1}$$

where $\sqrt{\langle V^2 \rangle}$ is the root-mean-square of the velocity of the mercury atoms and

$$\omega = \frac{2\pi c}{\lambda} = (2\pi)\left(\frac{3.0 \times 10^{8}}{2.537 \times 10^{-7}}\right) = 7.43 \times 10^{15} \text{sec}^{-1}.$$

Using the equipartition principle we obtain

$$\frac{m}{2} \langle V^2 \rangle = \frac{kT}{2}$$

or

$$\sqrt{\langle V^2 \rangle} = \sqrt{\frac{kT}{m}} = c \sqrt{\frac{kT}{mc^2}} = c \sqrt{\frac{(1.38 \times 10^{-16})(3 \times 10^{2})}{202(1.66 \times 10^{-24})(9 \times 10^{20})}}$$

$$= 3.7 \times 10^{-7} c.$$

Therefore we have for (1)

$$\Delta\omega_D = 3.7 \times 10^{-7} \times \omega = (3.7 \times 10^{-7})(7.42 \times 10^{15}) \simeq 3 \times 10^{9}/\text{sec.}$$

The natural width is defined through the uncertainty principle

$$\Delta\omega_N = \frac{1}{\tau}$$

where τ is the mean time that such a transition takes place. For electric dipole transition, $\tau \approx 10^{-8}$ sec or $\Delta\omega_N \approx 10^{8}\text{sec}^{-1} << \Delta\omega_D$. The total power radiated is

$$W = n\hbar\omega = \frac{n\hbar c}{\lambda} \quad \text{where } n = \text{number of photons per}$$

Therefore

$$W = 10^{18} \times \frac{1973}{400} = 5 \times 10^{18} \text{ eV} \sim 1 \text{ joule.}$$

(I-4)

a. $1s^2\ 2s^2\ 2p^6\ 3s^2\ 3p^6\ 4s^1$.

b. $L = 0,\ S = 1/2,\ J = 1/2$.

c.

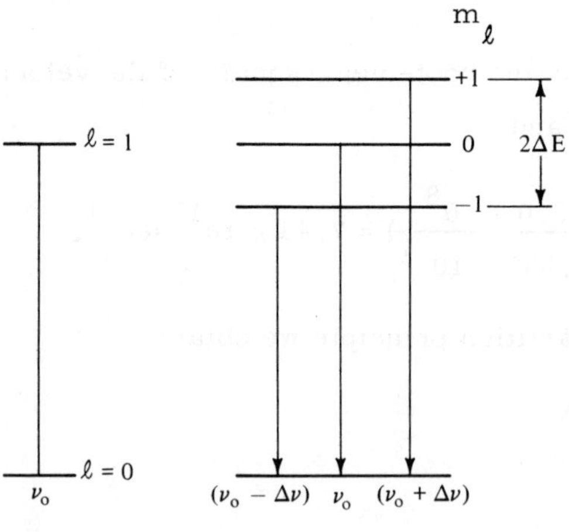

For normal Zeeman effect we have to consider only the energy split due to different values of m_ℓ, the z-component of the orbital angular momentum. For ground state $m_\ell = 0$, and for the first excited state (4p) $m_\ell = -1, 0, +1$. The energy levels and the allowed transitions are shown in the above figure. The energy shift ΔE in a magnetic field B is

$$\Delta E = \frac{\pm eh}{4\pi mc} B.$$

The selection rule for transitions between magnetic sublevels is $\Delta m_\ell = 0$ or ± 1. Thus

$$\Delta m_\ell = +1: \quad h\nu = h\nu_0 + \frac{eh}{4\pi mc} B$$

$$\Delta m_\ell = 0: \quad h\nu = h\nu_0$$

$$\Delta m_\ell = -1: \quad h\nu = h\nu_0 - \frac{eh}{4\pi mc} B.$$

(I-5)

a. Since ψ has to be normalized, we have the condition

$$\int_{-a}^{a} \psi^2 dx = 1$$

or

$$C^2 \int_{-a}^{a} (\cos^2 (\frac{\pi x}{2a}) + \sin^2 (\frac{3\pi x}{a}) + \frac{1}{16} \cos^2 (\frac{3\pi x}{2a}))dx = 1$$

where we have neglected all the cross terms which vanish after integration. From the above expression it is straightforward to carry out the integration. We find

$$2aC^2(\frac{1}{2} + \frac{1}{2} + \frac{1}{32}) = 1$$

or

$$C = 4\sqrt{\frac{1}{33a}}.$$

b. Substituting the three terms in ψ separately into the Schröedinger's equation we obtain

$$-\frac{\hbar^2}{2m}\frac{d^2}{dx^2}\cos\frac{\pi x}{2a} = \frac{\hbar^2}{2m}\frac{\pi^2}{4a^2}\cos\frac{\pi x}{2a} = \frac{\pi^2\hbar^2}{8ma^2}\cos\frac{\pi x}{2a} \equiv E_1 \cos\frac{\pi x}{2a},$$

$$-\frac{\hbar^2}{2m}\frac{d^2}{dx^2}\sin\frac{3\pi x}{a} = \frac{\hbar^2}{2m}\frac{9\pi^2}{a^2}\sin\frac{3\pi x}{a} \equiv E_2 \sin\frac{3\pi x}{a},$$

and

$$-\frac{\hbar^2}{2m}\frac{d^2}{dx^2}\cos\frac{3\pi x}{2a} = \frac{9\hbar^2\pi^2}{8ma^2}\cos\frac{3\pi x}{2a} \equiv E_3 \cos\frac{3\pi x}{a}.$$

We find these functions are the eigenstates of the system and E_1, E_2, and E_3 are the corresponding eigenvalues. Since the given wave function is a linear combination of the three eigenfunctions, the possible results of the measurement are

$$E_1 = \frac{\pi^2\hbar^2}{8ma^2}; \quad E_2 = \frac{9\pi^2\hbar^2}{2ma^2}; \text{ and } E_3 = \frac{9\pi^2\hbar^2}{8ma^2}.$$

The corresponding probability to measure each of them is proportional to the square of the weight of each wave function. We find

$$P_1 : P_2 : P_3 = 1 : 1 : \frac{1}{16}.$$

Since $P_1 + P_2 + P_3$ equals 1, we find

$$P_1 = \frac{16}{33}; \quad P_2 = \frac{16}{33} \text{ and } P_3 = \frac{1}{33}.$$

(II-1)

a. 939.5 MeV or 1.675×10^{-27} kg.

b. 6.626×10^{-27} erg-sec.

c. $\alpha = \frac{e^2}{\hbar c} = \frac{1}{137}$.

d. $\lambda_e = \frac{h}{m_e c} = \frac{2\pi\hbar c}{m_e c^2} = \frac{2\pi \times 1973}{5 \times 10^5} = 0.0248$ Å.

e. $r_o = \frac{e^2}{m_e c^2} = 2.82 \times 10^{-13}$ cm.

f. 10^{-9} seconds.

g. $\mu_P = \frac{2.7\, e\hbar}{2mc} = 8 \times 10^{-18} \frac{\text{MeV}}{\text{Gauss}}$.

h. $c\tau = 3.03 \times 10^{13}$ cm or $\tau = 10^{3}$ seconds.

i. $v = \alpha c = 2.2 \times 10^{8}$ cm/sec.

(II-2)

a. Substituting $-x$ for x into the time-independent Schröedinger equation

$$-\frac{\hbar^2}{2m}\frac{d^2\psi(x)}{dx^2} + V(x)\psi(x) = E\psi(x) \qquad (1)$$

we obtain

$$-\frac{\hbar^2}{2m}\frac{d^2\psi(-x)}{dx^2} + V(-x)\psi(-x) = E\psi(-x). \qquad (2)$$

Using $V(x) = V(-x)$, we can rewrite equation (2)

$$-\frac{\hbar^2}{2m}\frac{d^2\psi(-x)}{dx^2} + V(x)\psi(-x) = E\psi(-x) \qquad (3)$$

from which we see $\psi(-x)$ is also a solution of the Schröedinger equation corresponding to the same energy E. However we know that $\psi(x)$ is not degenerate; therefore $\psi(x)$ and $\psi(-x)$ must be linearly dependent. Since both wave functions are normalized, we obtain,

$$\psi(x) = \pm\, \psi(-x) \qquad (4)$$

where $\pm$ means $\psi(x)$ has either positive or negative parity.

b. The potential $V(x)$ satisfies the condition $V(x) = V(-x)$. Therefore as we proved in part a that the wave function has definite parity: $\psi(x) = \pm\, \psi(-x)$. Let

$$k_1 = \frac{n\pi}{2a} \text{ and } k_2 = \frac{1}{\hbar}\sqrt{(2mV_o - k_1^2\hbar^2)}$$

where n is an integer. Since the wave function must vanish at $x = \pm a$, $\psi(x)$ behaves like $\sin k_1(x + a)$ in the region $-a < x < -b$ and like $\pm \sin k_1(a - x)$ in the region $b < x < a$. Furthermore, ψ decreases exponentially according to

$$e^{-k_2(b+x)}, \qquad +(-)e^{-k_2(b-x)}$$

in the regions $-b < x < 0$ and $0 < x < b$, respectively, where +(-) means positive (negative) parity wave function. The ground state ψ_1 has positive (even) parity and the first excited state ψ_2 has negative (odd) parity. The energy for the nth state is

$$E = \frac{n^2\pi^2\hbar^2}{8ma^2} \qquad n = 1, 2, \ldots$$

The wave functions ψ_1, ψ_2, and $(\psi_1 + \psi_2)$ at $t = 0$ are plotted below as functions of x:

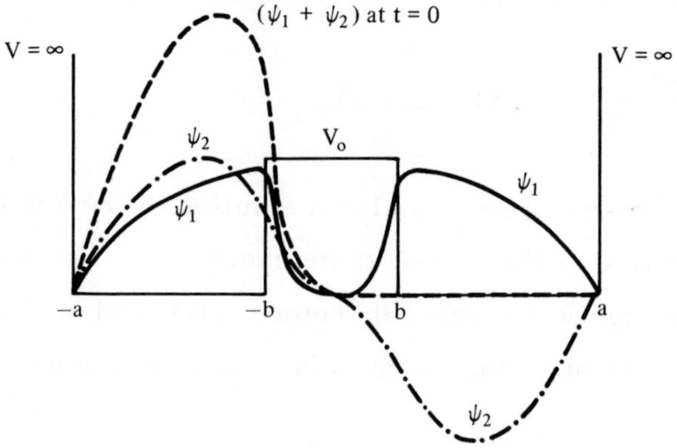

c. Let us construct a wave packet

$$\psi = \psi_1 e^{-iE_1t/\hbar} + \psi_2 e^{-iE_2t/\hbar}.$$

Since ψ_1 and ψ_2 are destructive in the right-hand well, constructive in the left-hand well, and negligible in the region $-b < x < b$, ψ is almost entirely in the left-hand well at $t = 0$. As time goes on, the phase difference between the two terms of ψ

changes as $(E_2 - E_1)t/\hbar$. Therefore the wave packet gradually moves to the right-hand well and then bounces back. The period of oscillation is

$$\frac{2\pi\hbar}{E_2 - E_1}.$$

(II-3)

a. $^1S_0$: The magnetic moment of the positron is pointing along the direction of its spin, while that of the electron is opposite to the direction of its spin. From classical electromagnetism we know that the potential energy of the system of two magnetic dipoles is lower when the two dipoles are in the same direction rather than when the dipoles are in the opposite direction. Therefore the spin of the positron is opposite to the direction of the electron for the ground state.

b. The reduced mass of positronium is $\mu = (1/2)m_e$. Therefore the binding energy = (1/2) (binding energy of H) = 6.8 eV.

c. $E_\gamma = (2m_e - \text{(binding energy)})/2 \sim 0.5$ MeV. Because of conservation of momentum and angular momentum, the two gamma rays are in the opposite direction with angular momenta antiparallel to each other.

(II-4)

a.

$$\langle \ell', m' | [L_+, L_-] | \ell, m \rangle$$

$$= \langle \ell', m' | [M_x + iM_y, M_x - iM_y] | \ell, m \rangle$$

$$= \langle \ell', m' | 2(iM_y M_x - iM_x M_y) | \ell, m \rangle = \langle \ell', m' | 2i[M_y, M_x] | \ell, m \rangle$$

$$= \langle \ell', m' | 2\hbar M_z | \ell, m \rangle$$

$$= 2\hbar^2 m \delta(\ell - \ell')\delta(m - m').$$

b. In the case that x is an operator (matrix) the functional e^{ix}, $\cos x$, and $\sin x$ are all defined by their corresponding

$$e^{i\sigma_y\theta/2} = 1 + (i\sigma_y \frac{\theta}{2}) + \frac{(i\sigma_y\theta/2)^2}{2!} + \ldots$$

$$= 1 - \frac{(\sigma_y\theta/2)^2}{2!} + \frac{(\sigma_y\theta/2)^4}{4!} - \ldots + i((\sigma_y \frac{\theta}{2})$$

$$- \frac{(\sigma_y\theta/2)^3}{3!} + \frac{(\sigma_y\theta/2)^5}{5!} - \ldots$$

Using the relation that $\sigma_y^2 = 1$ we find

$$e^{i\sigma_y\theta/2} = 1 - \frac{(\theta/2)^2}{2!} + \frac{(\theta/2)^4}{4!} - \ldots$$

$$+ i\sigma_y(\frac{\theta}{2} - \frac{(\theta/2)^3}{3!} + \frac{(\theta/2)^5}{5!} \ldots)$$

$$= \cos\frac{\theta}{2} + i\sigma_y \sin\frac{\theta}{2}.$$

(II-5) The electric potential inside a uniformly charged shell is $V = e/a$ = constant, while the corresponding potential for a point charge is e/r. The energy shift is

$$\Delta E = \int \psi^*(\frac{e^2}{r} - \frac{e^2}{a})\psi r^2 dr d\Omega$$

$$= (\frac{1}{a_o})^3 4\int_o^a e^{-2r/a_o}(\frac{e^2}{r} - \frac{e^2}{a})r^2 dr$$

$$\sim \frac{2e^2a^2}{3a_o^3} = \frac{4}{3} \times \frac{e^2}{2a_o} \times (\frac{a}{a_o})^2 = \frac{4}{3} \times 13.6\text{eV} \times (\frac{10^{-13}}{0.5 \times 10^{-8}})^2$$

$$\sim 7 \times 10^{-9} \text{ eV}$$

from which we see the energy level has increased. Therefore the binding energy decreases.

(III-1) What usually determines whether a nucleus is stable or not is the ratio r, defined by

$$r = \frac{\text{electrostatic energy}}{\text{surface energy of the nucleus}} \propto \frac{Z^2e^2/A^{1/3}}{A^{2/3}}$$

$$= \text{constant}\ \frac{Z^2e^2}{A}$$

where we have used the relation: radius $\propto A^{1/3}$ and area of nucleus surface $\propto A^{2/3}$. When r is small, the nucleus is stable. The nucleus becomes unstable when $r \geq r_o$, where r_o corresponds to $Z_o \sim 80$, $A_o \sim 200$.
We find

$$r_o \equiv \text{constant} \times \frac{(Z_o)^2e^2}{A_o} \geq \text{constant} \times \frac{Z^2(2e)^2}{A} \equiv \text{constant} \times \frac{4Z^2}{A}e^2$$

Thus

$$\frac{4Z^2}{A} \leq \frac{Z_o^2}{A_o} = \frac{(80)^2}{200} = 32;\ \text{further}\ \frac{4Z^2}{A} \simeq \frac{4Z^2}{2Z} = 2Z.$$

So $2Z = 32$ or $Z = 16$. The heaviest stable nucleus would have $Z \simeq 16$, $A \simeq 32$.

(III-2) The Paschen-Back effect occurs when $\mu_N B$ is comparable to the hyperfine separation which is of the order of 10^{-6} eV. Therefore

$$B = \frac{10^{-6}}{\mu_N} = \frac{10^{-1}}{3 \times 10^{-12}} \simeq 3 \times 10^5 \text{ gauss.}$$

(III-3) K.E. $= m_n c^2 + m_n c^2 = 2m_n c^2$.

(III-4) The Bohr radius of the muon is proportional to $1/Z$ and the number of protons in the nucleus is proportional to Z. Therefore the probability of the muon inside the volume occupied by the nucleus is proportional to Z^4.

(III-5) $E_P \lesssim 10$ MeV, i.e., when the de Broglie wave length of the nucleon is not much smaller than $\hbar/m_\pi c$, where m_π is the mass of the pion ($J^P = 0^{-1}$).

(III-6) This problem is similar to the problem of a linear harmonic oscillator. The potential in the region $x > 0$ is the same as that of a L.H.O. Therefore we have the same Schröedinger equation

$$H\psi = -\frac{\hbar^2}{2m}\frac{d^2\psi}{dx^2} + \frac{1}{2}kx^2\psi = 0 \quad \text{for} \quad x > 0.$$

However the boundary condition becomes

$$\psi_n(x = 0) = 0.$$

For even n's, the wave functions of a L.H.O. are even functions of x. These wave functions do not vanish at $x = 0$ and therefore do not satisfy the above boundary condition. For odd n's, the wave functions are odd functions of x. Therefore

$$\psi_{2m+1}(x = 0) \equiv 0$$

for any integer m. We see the functions

$$\psi_{2m+1}(x)$$

satisfy the boundary condition and the Schröedinger equation; therefore these functions are acceptable solutions. The corresponding energies are

$$E_{2m+1} = (2m + \frac{3}{2})\hbar\omega \quad \text{for} \quad m = 0, 1, 2, 3, \ldots$$

(III-7) Absolute symmetries and conserved quantities:

a. C.P.T. = 1.

b. Lorentz invariance.

c. Conservation of probability (unitarity).

d. Conservation of the number of Fermi particles.

e. Conservation of angular momentum, energy, and linear momentum.

f. Baryon number and lepton number.

g. Charge.

Partially conserved symmetries:

a. Parity.

b. Charge conjugation.

c. Time reversal invariance.

d. Charge independence of nuclear force.

e. G-parity.

f. Strangeness, etc.

(III-8) The differential decay rate of the particle at time t is

$$dP(t) = -\left(\frac{dt}{t_1} + \frac{dt}{t_2}\right)$$

$$= -\frac{t_1 + t_2}{t_1 t_2}\, dt$$

from which we find the probability that the meson still exists at t is

$$P(t) = \exp - \left(\frac{t_1 + t_2}{t_1 t_2}\right) t.$$

The uncertainty in time is then

$$\frac{t_1 t_2}{t_1 + t_2}.$$

The uncertainty in energy according to the uncertainty principle is

$$\Delta E = \frac{\hbar(t_1 + t_2)}{t_1 t_2}$$

or

$$\Delta m = \frac{\hbar(t_1 + t_2)}{(t_1 t_2) c^2}.$$

(III-9)

$$\text{Group velocity} = \frac{d\omega}{dk} = \frac{dE}{dp} = \frac{p}{m} = V.$$

$$\text{Phase velocity} = \lambda\nu = \frac{h}{mV}\nu = \frac{E}{mV} = \frac{c^2}{V}.$$

(III-10)

a. The 1s state vanishes only at infinity.

b. The 2s state vanishes at $r = 2a_o$ as well as at infinity.

c. The $2p(m = \pm 1)$ state vanishes at $\theta = 0, \pi$, and at $r = 0$ and $r = \infty$. The $2p(m = 0)$ state vanishes at $\theta = \pi/2, (3/2)\pi$, and at $r = 0$ and $r = \infty$. (Here m is the magnetic quantum number and θ = colatitude).

(III-11)

$$\epsilon = \frac{2\pi^2 me^4 Z^2}{h^2}$$

a. The effect of the relativity correction is to make the binding energy larger (energy level lower) because the mass of electrons becomes larger.

b. The screen effect reduces the effective Z; therefore the binding energy decreases.

c. As far as the charge distribution of the nucleus is uniform and spherically symmetric, the interaction potential energy between the electron and the nucleus is the same if we assume the electron is in a definite orbit according to Bohr atomic model. However, according to quantum mechanics, the electron has a finite probability of getting inside the nucleus. Therefore the binding energy is smaller if the nucleus has finite size. (See II-5.)

(III-12)

a. Only gravitational potential, $V \propto -1/r$.

b. $V \propto 1/r$ according to Coulomb's law.

c. The neutral atom acts like an electric dipole in the field of the ion. The potential energy $V \propto -1/r^2$.

d. The nuclear potential is of the form of $-e^{-kr}/r$.

e. Neutrons are neutral particles and have no electric dipoles.

The major contribution therefore comes from the interaction between the magnetic dipoles associated with the neutrons. The potential energy of two magnetic dipoles is of the form of $-2d_1 d_2/r^3$ with $d_1 = d_2 = d$ (magnetic dipole of a neutron).

(III-13) The total wave function can be written as a product of radical function R(r) and angular function $\psi(\theta, \varphi)$:

$$\psi(r, \theta, \varphi) = R(r)\psi(\theta, \varphi) \tag{1}$$

$\psi(\theta, \varphi)$ can be further decomposed into the angular momentum eigenstates as

$$\psi(\theta, \varphi) = \sum_{\ell, m} a_{\ell, m} Y_\ell^m(\theta, \varphi)$$

where $Y_\ell^m(\theta, \varphi)$ are the well-known spherical harmonics defined for ℓ and m integers and for $0 \leq m \leq \ell$. The eigenvalue of angular momentum corresponding to state

$Y_\ell^m(\theta, \varphi)$ is $\sqrt{\ell(\ell + 1)}\hbar$

The square of all

$|Y_\ell^m(\theta, \varphi)|^2$ except $|Y_o^o(\theta, \varphi)|^2$

are functions of θ; so are all the interference terms.* We find the only state satisfying the condition that

$|\psi(\theta, \varphi)|^2$

---

*If the sum of a certain interference terms is independent of θ and φ, i.e.,

is independent of θ and φ is

$$\psi(\theta,\varphi) \equiv Y_0^0(\theta,\varphi),$$

i.e., $\ell = m = 0$. Therefore the only allowed value of angular momentum is zero. The system has to be in a S state.

(IV-1) The one-dimensional Schröedinger equation is

$$\frac{-\hbar^2}{2m}\frac{d^2\psi}{dx^2} = (E - V)\psi. \tag{1}$$

For particles in the left-hand region, $V = 0$; (1) becomes

$$\frac{d^2\psi}{dx^2} = -\frac{2mE}{\hbar^2}\psi$$

the general solution of which is

$$\psi_L = Ae^{ik_1x} + Be^{-ik_1x} \tag{2}$$

where

$$k_1 = \frac{\sqrt{2mE}}{\hbar}.$$

For particles in the middle region, $V = V_o$, where $V_o > E$. The corresponding solution of Schröedinger equation is

$$\psi_M = Ce^{k_2x} + De^{-k_2x} \tag{3}$$

$$\sum_{i,j} a_{ij}\psi_{\ell_i}^{m_i^*}(\theta,\varphi)\psi_{\ell_j}^{m_j}(\theta,\varphi) = \text{const.} = A$$

then by integrating over $d\Omega$ and using the relation

$$\int \psi_\ell^m(\theta,\varphi)\psi_{i'}^{m'}(\theta,\varphi)d\Omega = 0 \text{ for } \ell \neq \ell' \text{ or } m \neq m'$$

we find $4\pi A = 0$. Therefore these terms vanish.

where

$$k_2 = \frac{\sqrt{2m(V_o - E)}}{\hbar}.$$

For particles in the right-hand region, $V = V_1 < E$. Since the wave is traveling only in the positive x-direction, the solution of (1) is

$$\psi_R = Fe^{ik_3 x} \quad (4)$$

where

$$k_3 = \frac{\sqrt{2m(E - V_1)}}{\hbar}.$$

The boundary conditions require that,

$$Ae^{-ik_1 d/2} + Be^{ik_1 d/2} = Ce^{-k_2 d/2} + De^{k_2 d/2} \quad (5)$$

$$ik_1 Ae^{-ik_1 d/2} - iBk_1 e^{ik_1 d/2} = k_2 Ce^{-k_2 d/2} - k_2 De^{k_2 d/2} \quad (6)$$

$$Fe^{ik_3 d/2} = Ce^{k_2 d/2} + De^{-k_2 d/2} \quad (7)$$

$$ik_3 Fe^{ik_3 d/2} = k_2 Ce^{k_2 d/2} - k_2 De^{-k_2 d/2}. \quad (8)$$

From (5) and (6) we obtain

$$2ik_1 Ae^{-ik_1 d/2} = (ik_1 + k_2)Ce^{-k_2 d/2} + (ik_1 - k_2)De^{k_2 d/2}. \quad (9)$$

From (7) and (8) we obtain

$$(k_2 + ik_3)Fe^{ik_3 d/2} = 2k_2 Ce^{k_2 d/2} \quad (10)$$

and

$$(k_2 - ik_3)Fe^{ik_3 d/2} = 2k_2 De^{-k_2 d/2}. \quad (11)$$

Eliminating C and D between (9), (10), and (11) we obtain

$$\frac{2ik_1Ae^{-ik_1d/2}}{Fe^{ik_3d/2}} = \left(\frac{(ik_1 + k_2)e^{-k_2d}(k_2 + ik_3)}{2k_2} + \frac{(ik_1 - k_2)e^{k_2d}(k_2 - ik_3)}{2k_2}\right)$$

from which we find the transmission coefficient is

$$\frac{k_3|F|^2}{k_1|A|^2} = \frac{16k_1k_2^2k_3}{(k_2^2 - k_1k_3)^2(e^{-k_2d} - e^{k_2d})^2 + (k_1k_2 + k_2k_3)^2(e^{-k_2d} + e^{k_2d})^2}$$

$$= \frac{4\sqrt{E(V_o - E)}\sqrt{(E - V_1)}}{((V_o - E) - \sqrt{E(E - V_1)})^2(\sinh^2 k_2d) + (V_o - E) \cdot (\sqrt{E} - \sqrt{(E - V_1)})^2 \cosh^2 k}$$

(IV-2) Let

$$A = \begin{pmatrix} 3 & 2 \\ 2 & 0 \end{pmatrix}$$

To find the eigenvalues, we have to solve the consistent equation,

$$|A - \lambda I| = 0 \tag{1}$$

or

$$\begin{vmatrix} 3 - \lambda & 2 \\ 2 & -\lambda \end{vmatrix} = 0 \tag{2}$$

$$(3 - \lambda)\lambda + 4 = 0$$

$$(\lambda - 4)(\lambda + 1) = 0. \tag{3}$$

Therefore $\lambda = 4$ or -1 are the two eigenvalues of A. For $\lambda = 4$, from the definition of eigenvector we have

$$\begin{pmatrix} 3 - 4 & 2 \\ 2 & -4 \end{pmatrix}\begin{pmatrix} x_1 \\ x_2 \end{pmatrix} = 0 \tag{4}$$

$$-x_1 + 2x_2 = 0 \quad \text{or} \quad x_1 = 2x_2. \tag{5}$$

The normalization condition is

$$x_1^2 + x_2^2 = 1. \tag{6}$$

The eigenvector corresponding to eigenvalue 4 is

$$\begin{pmatrix} \frac{2}{\sqrt{5}} \\ \frac{1}{\sqrt{5}} \end{pmatrix}.$$

For $\lambda = -1$

$$\begin{pmatrix} 3+1 & 2 \\ 2 & +1 \end{pmatrix}\begin{pmatrix} y_1 \\ y_2 \end{pmatrix} = 0 \tag{7}$$

i.e.,

$$4y_1 + 2y_2 = 0 \quad \text{or} \quad y_1 = \frac{-1}{2} y_2 \tag{8}$$

$$y_1^2 + y_2^2 = 1. \tag{9}$$

The eigenvector corresponding to eigenvalue -1 is

$$\begin{pmatrix} -\frac{1}{\sqrt{5}} \\ \frac{2}{\sqrt{5}} \end{pmatrix}$$

(IV-3)

a. $\vec{C}$ is the probability current density which corresponds to the Poynting vector in the classical electromagnetism theory.

b.

$$\frac{d}{dt}\int_V \psi^* \psi \, dv = \int_V \left(\psi^* \frac{d\psi}{dt} + \frac{\partial \psi^*}{\partial t}\psi\right)dv$$

$$= \frac{i\hbar}{2m}\int_V (\psi^* \nabla^2 \psi - (\nabla^2 \psi^*)\psi)dv$$

$$= \frac{i\hbar}{2m}\int_V \mathrm{div}(\psi^* \,\mathrm{grad}\, \psi - (\mathrm{grad}\, \psi^*)\psi)dv$$

$$= -\int_V \nabla \cdot \vec{C}dv.$$

Therefore

$$\vec{C} = \frac{\hbar}{2im}(\psi^* \,\mathrm{grad}\, \psi - (\mathrm{grad}\, \psi^*)\psi).$$

(IV-4)

a. For the ground state the wave faction is

$$\psi_o = 2(\frac{1}{a_o})^{3/2} \exp(-\frac{r}{a_o})\frac{1}{\sqrt{4\pi}}$$

where $a_o = 0.5$ Å.

b. Probability of finding the electron between r and $r + dr$ is

$P(r)dr = |\psi_o|^2 dv$ where $dv = 4\pi r^2 dr$.

Therefore

$$P(r)dr = (2(\frac{1}{a_o})^{3/2} \exp(-\frac{r}{a_o}))^2 r^2 dr \tag{1}$$

c. For maximum $P(r)$, $\frac{dP(r)}{dr}$ vanishes. We get

$$\frac{d}{dr}(\exp(-\frac{2r}{a_o})r^2) = 0$$

or

$$2r - \frac{2r^2}{a_o} = 0$$

which leads to

$$r = a_o$$

i.e., $P(r = a_o)$ is maximum. The other solution $r = 0$ corresponds to the case $P(r)$ is minimum, which is zero.

(IV-5)

a. $R \sim \frac{2\hbar^2}{me^2} \sim 1\text{Å}$.

b. Using the relation $\lambda T = 0.29°\text{K-cm}$ we find

$\lambda = 10^{-1}$ cm $= 10^7\text{Å}$ for $T = 3°\text{K}$.

c. Energy gap = 1.1 eV.

d. About 200 MeV is released in fission of a single nucleus. Energy released ≈ the binding energy of the initial nucleus- the binding energy of the final nucleus.

e.

$$f = \frac{c}{\lambda} \sim \frac{3 \times 10^{10}}{7 \times 10^{-5}} \sim 4 \times 10^{14} \text{ sec}^{-1}.$$

f.

$$t = \frac{d}{c} = \frac{3 \times 10^{-13}}{3 \times 10^{10}} \sim 10^{-23} \text{ sec}$$

g.

$$\mu_e = 0.9 \times 10^{-27} \text{ joules/gauss.}$$

(IV-6)

a. The normalized wave function of the ground state between two perfectly rigid walls separated by a distance a is,

$$u^o(x) = \sqrt{\frac{2}{a}} \cos \frac{\pi x}{a}$$

which vanishes at $x = -a/2$ and at $x = a/2$. The eigenfunctions after the expansion of the potential walls are

$$v^n(x) = \sqrt{\frac{1}{a}} \cos \frac{n\pi x}{2a} \quad \text{for } n \text{ odd}$$

$$v^n(x) = \sqrt{\frac{1}{a}}\sin\frac{n\pi x}{2a} \quad \text{for } n \text{ even } (n \neq 0)$$

which vanish at $x = a$ and at $x = -a$. We can expand $u^o(x)$ in terms of the set of orthonormal functions v^n:

$$u^o(x) = \sum_{n=1}^{\infty} a_n v^n(x)$$

where

$$a_n = \int_{-a/2}^{a/2} u^o(x) v^n(x) dx$$

is the probability amplitude for the particle to be in state n. The probability that the particle is in the ground state is

$$P = a_o^2 = \left(\int_{-a/2}^{a/2}\sqrt{\frac{2}{a}}\cos\frac{\pi x}{a}\frac{1}{\sqrt{a}}\cos\frac{\pi x}{2a}dx\right)^2 = \frac{2}{a^2}\left(\int_{-a/2}^{a/2}\cos\frac{\pi x}{a}\cos\frac{\pi x}{2a}dx\right)^2$$

$$= \frac{64}{9\pi^2}.$$

b. The energy is conserved in the statistical sense that the initial energy $E(u^o)$ equals the sum of all the possible energy levels times the probability to measure each of them, i.e.,

$$E(u^o) = \sum_i E_i(v^i)P_i$$

where E_i is the eigenvalue corresponding to v^i and P_i is the corresponding probability. (Refer to problem (I-5).)

(V-1)

a. Correspondence principle: the classical theory is a limiting case of quantum mechanics as the Planck constant h becomes negligible compared with some typical quantity, e.g., the angular momentum involved in the problem. All the laws and predictions in quantum mechanics should reduce to those in classical theory in this case.

b. Dulong and Petit law: the molar heat capacity of any metal equals 3R or 6 cal/mole-deg.

c. Because initially all atoms of alkali metals are in the S-state which is the ground state. A 3S electron can absorb a photon only when the photon is in the principal series.

d. g is 2 for electron spin angular momentum and 1 for electron orbital angular momentum.

e. Compton scattering and photoelectric effect.

(V-2)

a. For an electron, the energy radiated is

$$\epsilon_e = 13.6\left(\frac{1}{1} - \frac{1}{n^2}\right)$$

$$= 10.2 \text{ eV} \quad \text{for} \quad n = 2.$$

The energy levels are proportional to the mass m which is 210 times heavier in the case of a muon. Therefore

$$\epsilon_\mu = 2142 \text{ eV}.$$

b. The first Bohr radius is inversely proportional to the mass m. Using $a_o = 0.5$ Å in the case of an electron, we find

$$a_\mu = 0.5 \frac{1}{210} \simeq 0.0023 \text{ Å}.$$

c. The velocity $v = a_n \omega_n$ is independent of the mass m because a_n is inversely proportional to m while ω_n is proportional to m.

$$v = \frac{e^2}{n\hbar} = \frac{\alpha c}{n}$$

where α is the fine structure constant.

(V-3) Using the relation $\lambda_{max} T = 0.29$ cm-°K, we find the maximum intensity of radiation energy at T = 2000°K occurs at $\lambda_{max} = 1.45 \times 10^{-4}$ cm $= 1.45 \times 10^4$ Å. The corresponding photon energy is about 1 eV which is a factor of 2 lower than the energy of the sodium yellow line (2.07 eV). However the spectrum of the thermal radiation energy is

$$I(\lambda) = \frac{c_1}{\lambda^5} \frac{1}{e^{c_2/KT} - 1}$$

$$I(\lambda) = \frac{c_1}{\lambda^5} \frac{1}{e^{(1.44 \times 10^8)/\lambda T} - 1} \sim \frac{c_1}{\lambda^5} \times \frac{1}{e^{7.22 \times 10^4/\lambda} - 1}$$

where the second radiation constant $c_2 = 1.44 \times 10^{-2}$ m-°K = 1.44×10^8 Å-°K.

For $\lambda = 5890$Å

$$I(5890\ \text{Å}) \sim \frac{c_1}{(5890)^5} \frac{1}{e^{12.3} - 1}$$

which is small compared with $I(1.45 \times 10^4\ \text{Å})$, i.e.,

$$\frac{I(5890\ \text{Å})}{I(14,500\ \text{Å})} = \left(\frac{14,500}{5890}\right)^5 \frac{e^{4.98} - 1}{e^{12.3} - 1} \sim (2.46)^5 \frac{1.44 \times 10^2}{2.20 \times 10^5} = (88.6)6.54 \times 10^{-4}$$

$$= 0.0580,$$

but larger than $I(\lambda < 5900\ \text{Å})$. Since all the sodium atoms are in the ground state $3^2S_{1/2}$, the lowest available excited state is

$3^2P_{1/2, 3/2}$.

The transitions

$$3^2P_{1/2, 3/2} \rightarrow 3^2S_{1/2}$$

corresponds to the 5890-Å line. Since the transition rule of electric dipole transition is $\Delta L = \pm 1$, the next lowest excited state is the 4P state which corresponds to a line of wave length $\lambda = 3300$ Å. Since

$I(3300) \ll I(5900)$, this transition is negligible compared with the yellow line. Therefore we conclude that the yellow line is almost the only appreciable line at such low temperature.

(V-4) The Schröedinger equation is,

$$\frac{\hbar^2}{2m}\frac{d^2\psi_1}{dx^2} + E\psi_1 = 0 \quad \text{for} \quad x < 0 \tag{1}$$

$$\frac{\hbar^2}{2m}\frac{d^2\psi_2}{dx^2} + (E - V_o)\psi_2 = 0 \quad \text{for} \quad x > 0. \tag{2}$$

a. The solutions of (1) and (2) are

$$\psi_1 = Ae^{ik_1x} + Be^{-ik_1x} \quad \text{for} \quad x < 0 \tag{3}$$

$$\psi_2 = Ce^{ik_2x} + De^{-ik_2x} \quad \text{for} \quad x > 0 \tag{4}$$

where

$$k_1 = \frac{\sqrt{2mE}}{\hbar} \quad \text{and} \quad k_2 = \frac{\sqrt{2m(E - V_o)}}{\hbar}.$$

For particles incident from left, D vanishes. The boundary conditions are

$$\psi_1(0) = \psi_2(0) \quad \text{and} \quad \frac{d\psi_1(0)}{dx} = \frac{d\psi_2(0)}{dx} \tag{5}$$

which lead to

$$A + B = C$$

and

$$k_1(A - B) = k_2C$$

or

$$C = \frac{2k_1}{k_1 + k_2} A \tag{6}$$

$$B = \frac{k_1 - k_2}{k_1 + k_2} A . \tag{7}$$

From (3), (4), (6), and (7) we have

$$\begin{aligned} \psi_1 &= A(e^{ik_1x} + \frac{k_1 - k_2}{k_1 + k_2} e^{-ik_1x} \\ \psi_2 &= A\frac{2k_1}{k_1 + k_2} e^{ik_2x} . \end{aligned} \tag{8}$$

b. $v = \sqrt{2E/m}$, where v is the velocity of incident particles. In a box of unit cross section and length $\ell = v$, there is one particle. The density of incident particles is just A^2. Therefore we have

$$A^2 v = 1 \text{ or } A = (\frac{m}{2E})^{1/4} \tag{9}$$

c. For $E < V_o$,

$$k_2 = i\frac{\sqrt{2m(V_o - E)}}{\hbar} \equiv ik_3$$

(8) becomes

$$\begin{aligned} \psi_1 &= A(e^{ik_1x} + \frac{k_1 - ik_3}{k_1 + ik_3} e^{-ik_1x}) \\ \psi_2 &= 2A\frac{k_1}{k_1 + ik_3} e^{-k_3x} \end{aligned} \tag{10}$$

The incident particles cannot penetrate the potential wall and are reflected back. The characteristic distance particles can get into the $x > 0$ region is

$$\frac{1}{k_3} = \frac{\hbar}{\sqrt{2m(V_o - E)}} .$$

(V-5) Substituting the given wave function into the Schröedinger equation in the spherical coordinates,

$$-\frac{\hbar^2}{2m}\frac{1}{r^2}\frac{d}{dr}\left(r^2\frac{d\psi_k}{dr}\right) + V(r)\psi_k = E\psi_k \tag{1}$$

and carrying out the differentiation, we obtain

$$\frac{\hbar^2 k^2}{2m}\psi_k + V(r)\psi_k = E\psi_k, \tag{2}$$

where $V(r)$ is a constant which can be set to zero. From this we see

$$E = \frac{\hbar^2 k^2}{2m}.$$

The particle is a free particle.

(VI-1)

a. Probability current density

$$J = -\frac{i\hbar}{2m}\int_{\text{surface}} (\psi^* \operatorname{grad}\psi - (\operatorname{grad}\psi^*)\psi)\cdot dA$$

(Refer to problem (IV-3).

b. For $\psi = e^{ikx}$, $\psi^* = e^{-ikx}$ we obtain

$$J = -\frac{i\hbar}{2m}\int_S [e^{-ikx}(ik)e^{ikx} - (-ik)e^{-ikx}e^{ikx}]dA = \frac{\hbar k}{m}\int_S dA$$

$$= \frac{\hbar k}{m} = \text{propagating velocity of the wave.}$$

c. The current corresponding to a real wave function is identical to zero.

(VI-2) Using the equation of constant acceleration we find the typical transverse distance is

$$s = \Delta x + v_x t$$

where v_x is the transverse velocity: $v_x = \Delta P_x / m$

$$s = \Delta x + \frac{\Delta p_x}{m}\sqrt{\frac{2H}{g}}$$

$$\sim \Delta x + \frac{\hbar}{m}\,\frac{2H}{g}\,\frac{1}{\Delta x}, \qquad (1)$$

where we have used the relation $\Delta p_x \sim \frac{\hbar}{\Delta x}$ according to the uncertainty principle. From (1) we find the condition for minimum of s is

$$\frac{\Delta s}{\Delta x} = 1 + \frac{\hbar}{m}\sqrt{\frac{2H}{g}}\left[\frac{-1}{(\Delta x)^2}\right] = 0$$

where

$$(\Delta x)^2 = \frac{\hbar}{m}\sqrt{\frac{2H}{g}} = \sqrt{\frac{2\hbar^2 H}{m^2 g}} \quad \text{or} \quad \Delta x \approx \left(\frac{2\hbar^2 H}{m^2 g}\right)^{1/4}$$

Therefore

$$s \sim 2\left(\frac{2\hbar^2 H}{m^2 g}\right)^{1/4}$$

(VI-3)

a. Comparing the formula $\sqrt{\nu} = aZ - b$ with the relation

$$h\nu = E_f - E_i = (Z-1)^2 R_c h\left(1 - \frac{1}{2^2}\right) = \frac{2\pi^2 m e^4 (Z-1)^2}{h^2}\,\frac{3}{4}$$

we find

$$a = \frac{1}{2}\sqrt{3Rc} = \frac{\pi e^2}{2h}\sqrt{\frac{6m}{h}}$$

where R is the Rydberg constant.

b. In an atom with large Z, all the electrons except the 1S electrons are usually much farther from the nucleus than the 1S electrons. Thus the 1S electrons have energy levels almost as if the other electrons were not present. If we take $E(1S) \propto (Z - 1)^2$ and $E(2P) \propto (1/2^2)(Z - 1)^2$, we obtain

$$\nu = \frac{E(2P) - E(1S)}{h}$$

or $\nu \propto (Z - 1)^2$ which is a special case of the Moseley's law.

For optical frequencies we cannot neglect the interaction effect of other electrons since in these cases transitions of electrons in the higher states are involved.

(VI-4) Let the x-axis be along the first particle's direction. The four momentum of the two particles are

$$P_1 = (\gamma_1 m_1 v,\ 0,\ 0,\ i\gamma_1 m_1 c)$$

$$P_2 = (0,\ 0,\ 0,\ im_2 c)$$

where

$$\gamma_1 = (1 - \frac{v^2}{c^2})^{-1/2} \quad .$$

The rest mass of the resultant particle is

$$M = -(\sum_{i=1}^{4} (P_1 + P_2)_i \cdot (P_1 + P_2)_i)^{1/2} \frac{1}{c}$$

$$= (m_1^2 + m_2^2 + 2\gamma_1 m_1 m_2)^{1/2}$$

The velocity V can be calculated from P, the linear momentum of the resultant particle. Since momentum is conserved, we have

$$P \equiv \frac{MV}{\sqrt{1 - \frac{V^2}{c^2}}} = \gamma_1 m_1 v$$

or

$$\gamma_1^2 m_1^2 v^2 (c^2 - V^2) = M^2 V^2 c^2$$

from which we find

$$V = \frac{\gamma_1 m_1 vc}{\sqrt{M^2 c^2 + \gamma_1^2 m_1^2 v^2}} .$$

(VI-5)

a. The torque acting on an electron is

$$\vec{\tau} = \vec{\mu} \times \vec{B}$$

$$= - \frac{e}{2mc} g(\vec{s} \times \vec{B})$$

or

$$\tau = \frac{egsB}{2mc} \quad (\text{for } \vec{s} \perp \vec{B}).$$

The spin precession frequency equals the ratio of the torque to the angular momentum of the electron,

$$\vec{\Omega} = \frac{\vec{s} \times \vec{\tau}}{s^2}$$

or

$$\Omega = \frac{egB}{2mc}$$

along the direction of B-field.

b. The cyclotron frequency is

$$\omega = \frac{eg}{2mc} \sim 8.8 \times 10^6 \text{ rad-sec}^{-1}\text{-gauss}^{-1} \quad \text{for small } \alpha.$$

c.

$$\varphi = \Omega t = \Omega \frac{\pi R}{v}$$

$$= \pi \frac{egB}{2mc} \frac{mc}{eB} = \frac{\pi g}{2}$$

$$= \pi + \frac{\alpha}{2}.$$

Therefore a measurement of φ determines the value of α. This is the principle of the famous g - 2 experiment which measures the small deviation of the g-factor of the muon from 2.

(VI-6)

a. Since the electron moves in a circular orbit, the condition of equilibrium is:

$$\frac{e^2}{r^2} = m\frac{v^2}{r} \tag{1}$$

where v is the velocity of the electron, and r is the radius of the orbit.

The Bohr quantum condition is:

Angular momentum of the electron $\equiv mvr = n\hbar$ for $n = 1, 2, 3, \ldots$ (2)

Eliminating v with (1) and (2) we obtain

$$r = \frac{n^2\hbar^2}{me^2}. \tag{3}$$

The potential energy V is

$$V = -\frac{e^2}{r}. \tag{4}$$

The kinetic energy of the electron is

$$T = \frac{1}{2}mv^2 = \frac{e^2}{2r}. \tag{5}$$

The total energy E, is then

$$E = -\frac{e^2}{2r} = -\frac{me^4}{2n^2\hbar^2} \quad n = 1, 2, 3, \ldots$$

b.

$$E(n = 4) - E(n = 2) = \frac{me^4}{2\hbar^2}\left(\frac{1}{2^2} - \frac{1}{4^2}\right) = 13.6 \text{ eV} \times \frac{3}{16} = 2.55 \text{ eV}.$$

(VII-1)

a. The Schröedinger equations are

$$-\frac{\hbar^2}{2m}\frac{d^2\psi(x)}{dx^2} + V_o\psi(x) = E\psi(x) \quad \text{for} \quad x > 0$$

$$-\frac{\hbar^2}{2m}\frac{d^2\psi(x)}{dx^2} = E\psi(x) \quad \text{for} \quad x < 0$$

where $V_o = 20$ eV; $E = 10$ eV.

b.

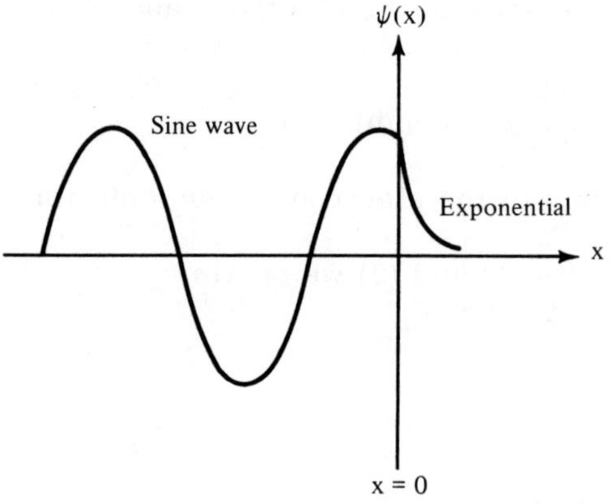

c. For $x < 0$

$$\lambda = \frac{h}{p} = \frac{h}{\sqrt{2mE}} = \frac{2\pi\hbar c}{\sqrt{(2mc^2)E}}$$

Using $m = 0.5$ MeV/c^2, $\hbar c = 1.973 \times 10^{-11}$ MeV-cm, and $E = 10$ eV, we find

$$\lambda = \frac{1.973 \times 2\pi \times 10^{-11}}{3.1 \times 10^{-3}} = 4 \times 10^{-8} \text{ cm}$$

where

$$\sqrt{(2mc)^2 E} = ((1.02 \text{ MeV})10^{-5} \text{ MeV})^{1/2} \approx 3.2 \times 10^{-3} \text{ MeV}.$$

d. The boundary conditions are that $\psi(x)$ and $\frac{d\psi(x)}{dx}$ are continuous at $x = 0$.

e. The probability of finding the electron at some positive value of x is proportional to

$\exp - 2x/\lambda$.

(VII-2)

a. Energy: $E_n = (n + \frac{1}{2})h\nu_o \qquad n = 0, 1, 2, \ldots$

Parity: $P = (-1)^n$.

b. Energy: $E_n = E_n^x + E_n^y + E_n^z$

$$= (n_x + n_y + n_z + \frac{3}{2})h\nu_o.$$

Parity: $P = (-1)^{n_x+n_y+n_z}$

Therefore the lowest four distinct groups of energy levels are:

$n_x + n_y + n_z$	E_n	Parity	n_x	n_y	n_z	ℓ	Degeneracy $= \sum (2\ell + 1)$
0	$\frac{3}{2}h\nu_o$	+1	0	0	0	0	1
1	$\frac{5}{2}h\nu_o$	-1	1	0	0	1	3
			0	1	0		
			0	0	1		
2	$\frac{7}{2}h\nu_o$	+1	1	1	0	2, 0	6
			0	1	1		
			1	0	1		
			2	0	0		
			0	2	0		
			0	0	2		

3	$\frac{9}{2}h\nu_0$	-1	1	1	1	3, 1	10
			2	1	0		
			1	2	0		
			2	0	1		
			1	0	2		
			0	2	1		
			0	1	2		
			3	0	0		
			0	3	0		
			0	0	3		

c. Using the relations: parity = $(-1)^{\ell}$ and number of states = $2\ell + 1$, we find the values of ℓ associated with each group as listed above in b.

(VII-3)

a. Since there are 6 2P-states, the number of ways for the first 2P electron is 6, while the number of states available for the second electron is only 5 because of the limitation of Pauli's principle. Therefore we have 30 ways of arrangement for the two 2P electrons. However, since the electrons are indistinguishable, the actual different ways of arrangement is $30/2 = 15$.

b. Let S and L be the total spin and orbital angular momentum of the two electrons. The total wave function of the two electrons, ψ, can be decomposed as a product of spin wave function $\psi_S(S_1, S_2)$ and orbital angular momentum wave function $\psi_L(r_1, r_2)$. When we interchange electron one and electron two, we find

$$\psi_L(r_1, r_2) = (-1)^L \psi_L(r_2, r_1)$$

and

$$\psi_S(S_1, S_2) = (-1)^{S+1} \psi_S(S_2, S_1).$$

Therefore

$\psi(1,\ 2) = (-1)^{L+S+1}\psi(2,\ 1).$ (1)

But according to Fermi statistics, the wave function of the two electrons must be antisymmetric, i.e.,

$\psi(1,\ 2) = (-1)\psi(2,\ 1).$ (2)

From (1) and (2) we find

$S + L = \text{even}.$ (3)

Since

$\vec{S} = \vec{S}_1 + \vec{S}_2$, S can be 0 or 1

and since

$\vec{L} = \vec{\ell}_1 + \vec{\ell}_2$, L can be 0, 1, or 2.

It follows that the possible ways of arrangement consistent with (3) are:

S	L	J	Multiplicity of states = 2J + 1
0	0	0	1
0	2	2	5
1	1	0	1
1	1	1	3
1	1	2	5

Total number of states = 15

(VII-4)

a. $1s^2\ 2s^2\ 2p^6\ 3s^1$.

b. $3\,^2S_{1/2}$.

c. $3\,^2P_{1/2}$, $3\,^2P_{3/2}$.

d. Fine structure splittings are caused by spin-orbit interactions, i.e., the interaction between the electron spin and the induction field B due to the nucleus in the rest frame of the electron.

e. The energy level of higher j is higher.

f. Since the Coulomb potential is $V \propto 1/r$, $dV/dr \propto r^{-2}$. The interaction energy is proportional to

$$\vec{S} \cdot \vec{L} \frac{1}{r}\frac{dV}{dr} \propto r^{-3} \vec{S} \cdot \vec{L} \propto r^{-3}$$

for fixed value of L. Therefore we find $n = -3$.

(VII-5)

a. $[L_x, L_y] = i\hbar L_z$
$[L_y, L_z] = i\hbar L_x$
$[L_z, L_x] = i\hbar L_y$, and
$[L_i, L^2] = 0 \quad i = x, y, z$

b.
$$\begin{aligned} L^2\varphi &= L^2(L_x + iL_y)\psi_{\ell m} \\ &= (L_x + iL_y)L^2\psi_{\ell m} \\ &= (L_x + iL_y)\hbar^2 \ell(\ell + 1)\psi_{\ell m} \\ &= \hbar^2 \ell(\ell + 1)\varphi. \end{aligned}$$

Thus φ is an eigenstate of L^2.

$$\begin{aligned} L_z\varphi &= L_z(L_x + iL_y)\psi_{\ell m} \\ &= (i\hbar L_y + L_x L_z + \hbar L_x + iL_y L_z)\psi_{\ell m} \\ &= (L_x + iL_y)(m\hbar + \hbar)\psi_{\ell m} \\ &= (m + 1)\hbar(L_x + iL_y)\psi_{\ell m} = (m + 1)\hbar\varphi \end{aligned}$$

so that φ is also an eigenstate of L_z.

c. Let $L_x\Psi_{oo} = \sum_{\ell, m} A_{\ell m}\psi_{\ell m} \quad m = 0, 1, 2, \ldots, \ell$

$$\ell = 0, 1, 2, \ldots \tag{1}$$

where $A_{\ell, m}$ are constant coefficients. Multiplying L^2 on both sides of (1) we get

$$L^2 L_x \Psi_{oo} = \hbar^2 \sum_{\ell, m} A_{\ell m} \ell(\ell + 1)\psi_{\ell m}. \tag{2}$$

The left side vanishes because L^2 commutes with L_x and $L^2\Psi_{oo} = 0$. Since the $\psi_{\ell m}$ are orthonormal functions, each term on the right side of (2) must vanish, i.e.,

$$A_{\ell m} = 0 \text{ unless } \ell = 0.$$

Therefore we find

$$L_x\Psi_{oo} = A_{oo}\Psi_{oo}$$

and similarly we can prove

$$L_y\Psi_{oo} = B_{oo}\Psi_{oo}.$$

(VIII-1)

a. ${}^2P_{1/2}$, since there is one unpaired electron.

b. ${}^1S_o$, since there are no unpaired electrons.

c. ${}^2P_{3/2}$, since there is one hole in the 2P shell.

d. In this case one 2P electron has jumped into 3S state leaving one hole in the 2P shell. Using Hund's rule we find

$$S = S_{2P} + S_{3S} = 1; \quad L = 1; \text{ and } J = 0.$$

Therefore we get ${}^3P_o$.

e. ${}^1S_o$.

(VIII-2)

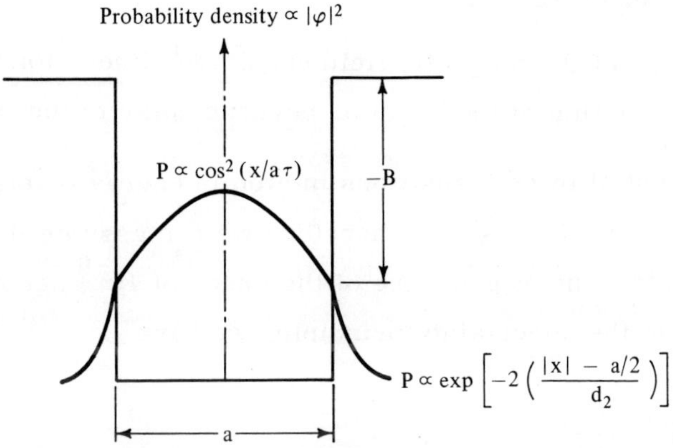

The wave function of the ground state in a potential well has only one maximum. The wavelength is approximately 2a inside the well. Outside the potential well, the probability density decreases exponentially, i.e.,

$$P \propto \exp(-2\,\frac{|x| - a/2}{d_2}).$$

For $x \equiv d >> a/2 + d_2$, the probability of finding the particle there becomes so small that it is negligible.

(VIII-3) The spin magnetic moment of an electron is $g_s \mu_B S$. The angular frequency of the Larmor precession is

$$\vec{\omega} = \frac{g_s \mu_B S}{\hbar}\vec{B} = \frac{e\vec{B}}{2mc} \qquad (g_s = 2, \quad S = \frac{1}{2}) \tag{1}$$

If the electrons change their spin directions by a large angle, say 180°, when they pass through the region between A and B, then the electrons will not be able to get through B since the deflection forces in A and B are in the same direction. Therefore we have the condition

$$\theta \sim \pi = \omega t = \frac{eBt}{2mc} \tag{2}$$

from which we find

$$B \sim \frac{2mc\pi}{et} = \frac{\pi}{8.8 \times 10^6 \times 10^{-6}} \sim 0.36 \text{ Gauss.}$$

The direction of the magnetic field should be either along the direction of motion of the beam or perpendicular to the x-z plane.

(VIII-4) Most allowed transitions in atomic energy levels take place in approximately 10^{-9} sec. Therefore we can assume the life time of the 2p state of hydrogen to be of the order of 10^{-9} second. According to the uncertainty principle, we have

$$\Delta E \sim \frac{\hbar}{\tau} \sim \frac{6.6 \times 10^{-16}}{10^{-9}}$$

$\sim 6.6 \times 10^{-7}$ eV

where

$$\hbar = \frac{1.05 \times 10^{-27} \text{ erg - sec}}{1.60 \times 10^{-12} \text{ erg/eV}} \simeq 6.6 \times 10^{-16} \text{ eV - sec.} \quad (1)$$

However there are two levels with $n = 2$ and $J = 1/2$, i.e., the levels $\ell = 0$ and $\ell = 1$. The $\ell = 0$ state is meta-stable; therefore its energy level can be measured much better than that in (1), while the energy level of the $\ell = 1$ state is more uncertain. The energy difference between the $\ell = 0$ level and the $\ell = 1$ level is so small that the frequency is in the microwave radio range. Since measurements of radio frequencies can be made very accurately, it is possible to measure the mean frequency of quanta absorbed in transitions between these two levels or equivalently the mean energy difference between them with great accuracy. Using the relation

$$\overline{E(\ell = 1)} = \overline{(E(\ell = 1) - E(\ell = 0))} + E(\ell = 0)$$

we can determine the <u>mean</u> energy level of $\ell = 1$ state better than that stated in (1).

(VIII-5)

a. $a_o = \frac{\hbar^2}{me^2} = 0.53\text{Å} = 5.3 \times 10^{-9}$ cm.

b. $r_o = \frac{\hbar^2}{mze^2} = 6.6 \times 10^{-3}\text{Å} = 6.6 \times 10^{-11}$ cm.

c. $\lambda\!\!\!^{-} = \frac{\hbar}{mc} = 3.8 \times 10^{-11}$ cm; Note $\lambda = \frac{h}{mc} = 2.42 \times 10^{-10}$ cm.

d. $\lambda\!\!\!^{-} = \frac{\hbar}{m_\pi c} = 1.4 \times 10^{-13}$ cm; Note $\lambda = \frac{h}{m_\pi c} = 8.88 \times 10^{-13}$ cm.

e. $\lambda\!\!\!^{-} = \frac{\hbar}{\sqrt{2m_n T}} = \frac{1973}{\sqrt{2 \times 10^9 \times 10^4}} \sim 0.44 \times 10^{-3}\text{Å} = 44 \times 10^{-13}$ cm.

f. $\lambda\!\!\!^{-} = \frac{\hbar}{P} \sim \frac{1973}{10^{10}} = 1.973 \times 10^{-7}\,\text{Å} = 1.973 \times 10^{-15}\,\text{cm}.$

g. $\lambda\!\!\!^{-} = \frac{\hbar}{m_\nu c} \gtrsim 10^{-3}\,\text{Å} = 10^{-11}\,\text{cm}$ for muon-type neutrino with

$m_\nu < 1.2\text{MeV}$,

and

$\lambda\!\!\!^{-} = \frac{\hbar}{m_\nu c} \gtrsim 30\text{Å} = 3 \times 10^{-7}\,\text{cm}$ for electron-type neutrino with

$m_\nu < 60\text{eV}$.

h. $r = 1.3 \times 10^{-13}\,\text{cm} \times \sqrt[3]{A}$ and $\sqrt[3]{A} \sim 6$. Therefore $r \sim 8 \times 10^{-13}\,\text{cm}$.

(VIII-6)

a. Let P be the parity operator which reflects all coordinates through the origin. Assuming all states have definite parity, we find

$$\begin{aligned}
\langle\vec{p}\rangle &= \langle \psi, \vec{p}\psi \rangle = \langle P\psi, \vec{p}P\psi\rangle = \langle PP\Psi, P\vec{p}P\Psi\rangle \\
&= \langle \psi, P\vec{p}P\psi \rangle \\
&= \langle \psi, -\vec{p}\psi \rangle = -\langle \Psi, \vec{p}\Psi \rangle \\
&= -\langle\vec{p}\rangle, \quad \text{where } PP = 1
\end{aligned}$$

Therefore

$\langle\vec{p}\rangle \equiv 0.$

b. If all the eigenstates corresponding to an eigenvalue E_i have definite and the same parity, the expectation value of the position vector $\vec{r}$ must vanish.

(VIII-7)

a. Assuming the L-S coupling prevails for light atoms, we find the following rules for electric dipole transition.

i. Parity must change

ii. $\Delta L = \pm 1$, $\Delta M = \pm 1, 0$

iii. $\Delta S = 0$,

iv. $\Delta J = 0, \pm 1$ (but not $J_i = 0 \rightarrow J_f = 0$).

b. If we assume J-J coupling for heavy atoms we have the following rules:

i. Only one electron jumps at a time.

ii. $\Delta J = 0, \pm 1$ and for the jumping electron, $\Delta \ell = \pm 1$ and $\Delta j = 0$ or ± 1.

iii. Parity changes.

c. In nuclei we have the restriction in the change of the nuclei spin: $\Delta I = \pm 1, 0$. Parity must also change (but not $I_i = 0 \rightarrow I_f = 0$).

(VIII-8) Hyperfine structure is due to the interaction between the nuclear magnetic dipole and the magnetic field B_e produced by the electron orbit current. The interaction energy is $\vec{\mu}_n \cdot \vec{B}_e$ where $\mu_n = e\hbar/2Mc$ and therefore is proportional to the inverse of M.

Fine structure is due to the interaction between, μ_e, the electron magnetic dipole and the Coulomb field of the nucleus, $\vec{E}$. The interaction energy is $\vec{\mu}_e \cdot \vec{E}$ where $\mu_e = e\hbar/2mc$ and therefore is proportional to the inverse of m. Furthermore $\vec{E}$ increases with increasing Z near $r = 0$. If we neglect the screen effect, we find $E \propto Z$. Therefore we find

$$\frac{hfS}{fS} \sim \frac{1/M}{Z/M} = \frac{m}{MZ}.$$

(VIII-9)

$$f(x) = \sum_{n=0}^{\infty} \frac{x^n}{(n!)^p} = \sum_{n=0}^{\infty} a_n$$

where

$$a_n = \frac{x^n}{(n!)^p}$$

$$\frac{a_{n+1}}{a_n} = \frac{x}{(n+1)^p} \rightarrow 0 \text{ as } n \rightarrow \infty .$$

Therefore f(x) converges for any finite value of x but goes to infinity faster than any finite power of x as x approaches infinity.

b. The integral diverges, therefore f(x) is not an acceptable solution.

(IX-1)

$$K_\alpha = \text{kinetic energy of alpha particle} = \frac{P_\alpha^2}{2M_\alpha} = \frac{(Q_\alpha BR)^2}{2M_\alpha}$$

$$K_D = \text{kinetic energy of deuteron} = \frac{P_D^2}{2M_D} = \frac{(Q_D BR)^2}{2M_D}$$

from which we find

$$\frac{K_\alpha}{K_D} = \frac{M_D}{M_\alpha}\frac{Q_\alpha^2}{Q_D^2} = \frac{2}{4} \times \frac{2^2}{1} = 2.$$

Therefore

$$K_\alpha = 2 \times K_D = 32\text{MeV}.$$

(IX-2) 10^{15} gm/cm^3, which can be estimated by multiplying the density of ordinary matter by the ratio of the atom volume to the volume occupied by the nucleus, i.e.,

$$\rho \approx 1 \frac{\text{gm}}{\text{cm}^3} \times \left(\frac{r_a}{r_n}\right)^3 = \left(\frac{10^{-8}}{10^{-13}}\right)^3 = 10^{15} \frac{\text{gm}}{\text{cm}^3}.$$

(IX-3) 4, since the Coulomb scattering is proportional to Z^2.

(IX-4) $mc^2 = 2 \times (3 \times 10^8)^2 = 1.8 \times 10^{17}$ joules.

(IX-5) 6.8 eV, since the reduced mass is one half of the electron mass.

(IX-6) There are six transitions, two each of three different energies. Therefore there are three spectral lines, corresponding to $\Delta m_J = +1,\ 0,$ and -1.

(IX-7) Using uncertainty principle we find $p \approx \hbar/R$. Therefore

$$T = \frac{\hbar^2}{2R^2 m}.$$

(IX-8) Using the relation $\hbar c/\lambda\!\!\!^{-} = 25$ keV, we find $\lambda\!\!\!^{-} = 0.08$Å.

(IX-9) $(1/2)^{1/3}$, since $P_{max} \propto (V)^{-1/3}$.

(IX-10) $2.8\,\mu_N$ or 1.4×10^{-23} erg/gauss, where μ_N = nuclear magneton.

(IX-11) ^1S_o.

(IX-12) K.E. increases by a factor of α^2; P.E. decreases by a factor of α^2, because

$$\text{K.E.} = <\psi(\alpha x),\ \frac{\hbar^2}{2m}\frac{d^2}{dx^2}\psi(\alpha x)> = \alpha^2 <\psi(\alpha x),\ \frac{\hbar^2}{2m}\frac{d^2}{d(\alpha x)^2}\psi(\alpha x)>$$

and

$$\text{P.E.} = <\psi(\alpha x),\ \frac{k}{2}x^2\psi(\alpha x)> = \frac{1}{\alpha^2}<\psi(\alpha x),\ \frac{k}{2}\alpha^2 x^2\psi(\alpha x)>$$

$$= \frac{1}{\alpha^2}<\psi(x),\ \frac{k}{2}x^2\psi(x)>.$$

(IX-13)

a. The current due to the ground state electron is

$$i = \frac{ev}{2\pi a}. \qquad (1)$$

The magnetic field at the center of a loop with radius a and current i is known to be

$$B = \frac{\mu_o i}{2a} = \frac{\mu_o ev}{4\pi a^2} = (10^{-7})\frac{(1.6 \times 10^{-19})(2.2 \times 10^6)}{(0.53 \times 10^{-10})^2} = 12.5\ \frac{\text{weber}}{m^2} =$$

1.25×10^5 gauss,

where $v = \alpha c = 2.2 \times 10^6$ m/sec and $a = 0.53 \times 10^{-10}$ m.

b. The energy difference between two states with proton spin up and with proton spin down is

$$\Delta E = 2B\mu_p = (2.5 \times 10^5)(3.15 \times 10^{-12}) = 7.88 \times 10^{-7}\text{eV}$$

where $\mu_p = 3.15 \times 10^{-12}$ eV/gauss.

c.

$$\omega = \frac{\Delta E}{\hbar} = \frac{7.88 \times 10^{-7} \times 3 \times 10^{18}}{1973} = 1.2 \times 10^9 \text{rad/sec.}$$

(IX-14)

a.

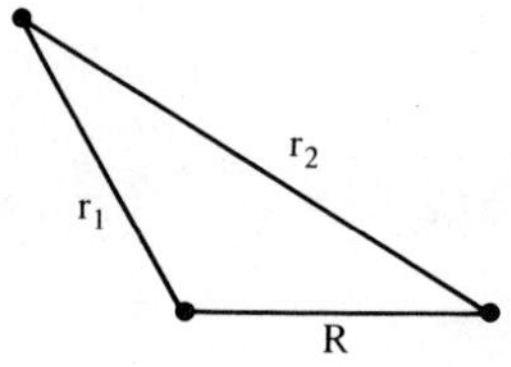

For $R = 0$, we have two positive charges at the origin. Therefore $E_o(0)$ should be four times the ground state energy of hydrogen.

$$E_o(0) = -4(R_H) = -54.4\text{eV}. \tag{1}$$

For $R \to \infty$ the second proton has no effect on the electron. The ground state for the electron is just the same as that in the hydrogen atom. Therefore

$$E_o(\infty) = -R_H = -13.6\text{eV}. \tag{2}$$

b. The total potential energy is

$$V(R) = E_o(R) + \frac{e^2}{R} \tag{3}$$

from which we find the equilibrium condition is

$$\frac{d}{dR}V(R) = 0. \tag{4}$$

Substituting (3) into (4), we get

$$-\frac{e^2}{R^2} + \frac{dE_o(R)}{dR} = 0 \tag{5}$$

or

$$R_{equ} = \frac{e}{\sqrt{\left(\frac{dE_o(R)}{dR}\right)_{R=R_{equ}}}} \tag{6}$$

c.

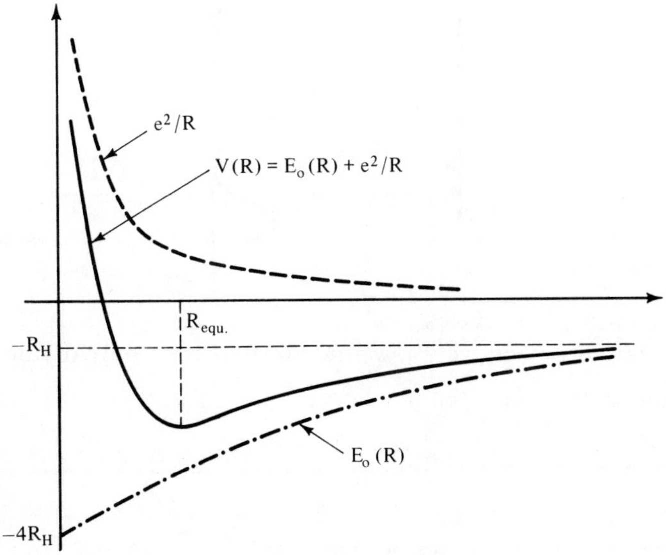

d. If the two protons are in low excited state, there will be two additional energy terms corresponding to the vibrational and rotational energy of protons:

$$U_{vib} = (n + \frac{1}{2})\hbar\omega_c \qquad n = 1, 2, 3, \ldots$$

where ω_c can be expressed in term of V(R), mass of proton and R_{equ}. The rotational energy is

$$U_{rot} \propto \frac{1}{M_p} J(J + 1) \qquad J = 1, 2, 3, \ldots$$

(IX-15)

a. The final energy of an electron is 37 eV. The de Broglie wavelength is therefore,

$$\lambdabar = \frac{\hbar}{p} = \frac{\hbar}{\sqrt{2mE}} = \frac{\hbar c}{\sqrt{2mc^2 E}} = \frac{1973}{\sqrt{1.02 \times 10^6 \times 37}}$$

$$= 0.32\text{Å}.$$

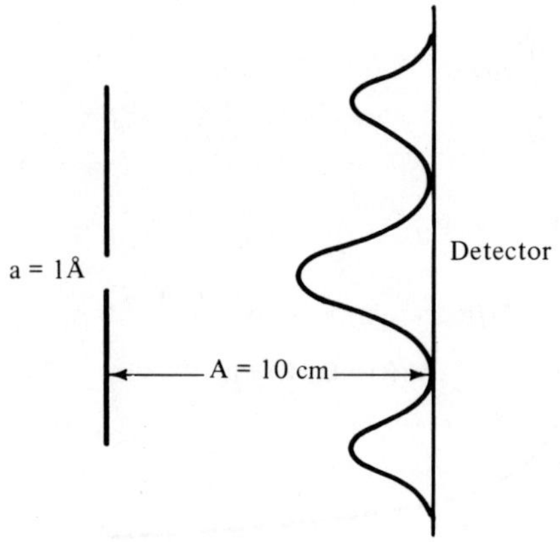

From classical optics we know the width of the central line of the diffraction pattern is

$$x \sim \frac{2\lambda}{a} A \sim 6.4 \text{ cm}$$

over which the electron can be recorded.

b. The intensity pattern becomes the double slit interference pattern.

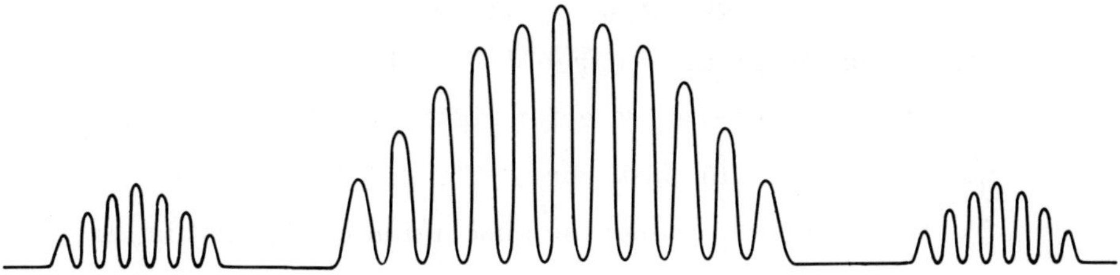

c. According to quantum mechanics, the wave function of a single electron can interfere with itself. Therefore we should observe the same pattern as that in b.

d. If there were such a detector and if one could select events with the electrons passing through the first slit, the single slit pattern would return. However, the uncertainty in the positions and momenta of the electrons and the detectors makes the second detector so unreliable (i.e., we can neither keep the second detector stationary at slit one nor do we know the incoming electrons are really parallel) that hardly any additional information can be obtained from the second detector. The double slit pattern persists. Refer to p. 161, Fundamental of Modern Physics, by Eisberg.

(X-1)

a. With $Z = 15$ and $m_\mu = 206\ m_e$ we find

$$E_{3\to 2} = Z^2 R_\infty (206)\left(\frac{1}{2^2} - \frac{1}{2^3}\right); \qquad R_\infty = 13.6\text{eV}$$

$$= 88\ \text{keV}.$$

b. The energy at which the lead K absorption edge occurs is

$$E_K = (Z-1)^2 R_\infty = (81^2)(13.6\ \text{eV}) = 89.3\ \text{keV}$$

The range of K X-rays therefore is 67--89.3 keV. When the μ-meson falls from the $n = 3$ level to the $n = 2$ level, it emits an X-ray which can excite one K-shell electron in the lead atom to higher states. Therefore the absorption coefficient is quite large for lead.

c. Let us do the following:

i. Set up the gamma source defined in a.

ii. Put a gamma-ray detector right behind a piece of lead foil, which serves as the absorber. We will detect a weak signal from the source because the absorption coefficient is large for $Z = 82$ element as we found in b.

iii. One at a time, replace the lead foil with some other absorbers of lower Z, i.e., $Z = 81, 80, 79, \ldots,$ all absorbers should have the same radiation lengths. Try to detect the X-ray as in part ii.

iv. Plot the observed X-ray intensity against the atomic number Z of the corresponding absorber used. At a certain point $Z = Z_1$, we will notice a sharp increase of the observed intensity as a function of Z, i.e., for absorbers of $Z < Z_1$, relatively much stronger signals are detected, while for absorbers of $Z > Z_1$ only weak signals are detected. We thus obtain the relation:

K absorption edge of element $Z_1 + 1 \geq E_{3-2} \geq$ K absorption edge of element Z_1. Here $Z_1 = 81$.

From which we can set the upper and the lower limit of the muon mass, i.e., $203\ m_e \leq m_\mu \leq 209\ m_e$.

If the mass of the muon were much larger, then Z_1 would be much larger than 82 and such element would not be found among the known stable elements and would not be easily available.

(X-2)

a. The energy scattered by the charge equals the energy radiated by it. Therefore

$$\frac{\text{Energy scattered}}{\text{Unit time}} \equiv \frac{dw}{dt} = \frac{2}{3}\frac{e^2 a^2}{c^3} \tag{1}$$

If the electric field at the location of the charge due to the incident radiation is E, then

$$a = \frac{F}{m} = -e\frac{E}{m} \tag{2}$$

From (1) and (2) we find

$$\frac{dw}{dt} = \frac{2e^4E^2}{3m^2c^3}. \tag{3}$$

The intensity of the incident radiation is

$$I = \frac{cE^2}{4\pi}. \tag{4}$$

The cross section σ_T by definition is

$$\sigma_T \equiv \frac{1}{I}\frac{dw}{dt} = \frac{8\pi}{3}\left(\frac{e^2}{mc^2}\right)^2 \tag{5}$$

where we have eliminated E with (3) and (4).

b. Thomson scattering is the nonrelativistic limit of Compton scattering. They are also equal in the forward scattering when the change in wavelength is small, i.e.,

$$\sigma_c = \sigma_T f(\cos\theta, E_\gamma)$$

with $f(1, E_\gamma \ll m_e c^2) \sim 1$.

c. Using the conservation laws of energy and momentum we find the following formula for Compton scattering:

$$E_{\text{scattered}} = \frac{E_{\text{incident}}}{\frac{E_{\text{incident}}}{m_o c^2}(1 - \cos\theta) + 1}. \tag{6}$$

For $\theta = 90°$, $m_o c^2 = m_e c^2 = 0.5$ MeV, and $E_{\text{incident}} = 0.5$ MeV, we get

$$E_{\text{scattered}} \approx 0.25 \text{ MeV}.$$

If it is scattered from a proton, $m_o c^2 \gg E_{incident}$

$E_{scattered} = E_{incident}$.

From (5) we see the cross section is proportional to $1/m^2$, therefore

$$\frac{\gamma + e^-}{\gamma + p^+} \sim \frac{m_p^2}{m_e^2} \sim 4 \times 10^6$$

when E, the energy of the gamma ray, is small.

(X-3) The energy of an oscillator with quantum number n is

$$E_n = (n + \frac{1}{2})\hbar\omega \qquad (1)$$

where ω is the angular velocity. The corresponding classical expression is

$$E_n = \frac{m}{2} A^2\omega^2, \qquad (2)$$

where A is the amplitude of the oscillator. From (1) and (2) we obtain

$$A^2 = (2n + 1)\frac{\hbar}{\omega m}. \qquad (3)$$

Substituting $a = A\omega^2$ into

$$\frac{dw}{dt} = \frac{2}{3}\frac{e^2a^2}{c^3}$$

we get

$$\frac{dw}{dt} = \frac{2}{3}\frac{e^2A^2\omega^4}{c^3} = \frac{2}{3}\frac{e^2\omega^3}{mc^3}(2n + 1)\hbar. \qquad (4)$$

The transition rule for an oscillator is $\Delta n = \pm 1$. When the electron has fallen from state n into the state $(n - 1)$, we say it is no longer in the n-state. Therefore the mean life is

$$\tau = \frac{E_n - E_{n-1}}{dw/dt}$$

$$= \frac{3mc^3}{2e^2\omega^2(2n+1)}$$

(X-4)

a. The Schröedinger equation for a particle inside a potential well is

$$-\frac{\hbar^2}{2m}\left(\frac{\partial^2\psi}{\partial x^2} + \frac{\partial^2\psi}{\partial y^2} + \frac{\partial^2\psi}{\partial z^2}\right) = E\psi .$$

Since the potential becomes infinite at the boundary, therefore $\psi(x, y, z) = 0$ at the boundary. The allowed values of total energy E of one proton (or neutron) is

$$E = \frac{\pi^2\hbar^2}{2ma^2}(n_x^2 + n_y^2 + n_z^2) = \frac{\pi^2\hbar^2 n^2}{2ma^2}$$

where n_x, n_y, and n_z are positive integers and a is the dimension of the potential well. Therefore we find
$dn/dE = ma^2/\pi^2\hbar^2 n$.

b. Since the total number of protons (or neutrons) is related to n by (n is positive and there are two spin states corresponding to each n)

$$N \text{ (or } Z) = \frac{1}{8}\left(\frac{4}{3}\pi n^3\right) \times 2 = \frac{\pi n^3}{3} \quad \text{or} \quad n^2 = \left(\frac{3N}{\pi}\right)^{2/3}$$

we find

$$E_f = 3^{2/3}\pi^{4/3}\frac{\hbar^2 N^{2/3}}{2ma^2} \quad \text{for a neutron} \qquad (1)$$

and similarly

$$E_f = 3^{2/3}\pi^{4/3}\frac{\hbar^2 Z^{2/3}}{2ma^2} \quad \text{for a proton} \tag{2}$$

c. Let $A/a^3 = \rho$, $N/A = \alpha_n$, and $Z/A = \alpha_p$. Note: $A = (N + Z)$. Since ρ, α_n, and α_p are constants, we find $N = \alpha_n A = \alpha_n \rho a^3$, and $Z = \alpha_p A = \alpha_p \rho a^3$, so that

$$\frac{N^{2/3}}{a^2} = (\alpha_n \rho)^{2/3} = \text{constant}$$

$$\frac{Z^{2/3}}{a^2} = (\alpha_p \rho)^{2/3} = \text{constant}$$

Therefore E_f in (1) or (2) is a constant.

d. We have to enhance the total potential acting on the protons. Thus the energy levels of protons are all elevated.

(X-5) If we assume the orbital angular momentum of the 3d electron of the Ti^{3+} irons is totally quenched, i.e., $\langle L_z \rangle = 0$, there is no contribution to the effective magnetic moment due to the orbital angular momentum. (Refer to p. 629, Introduction to Solid State Physics by Charles Kittle.) We are dealing with the normal Zeeman effect. The energy levels are split according to different values of m_s in the magnetic field. We get $g = 2$ and

$$N_{1/2} : N_{-1/2} = \exp\left(\frac{x}{kT}\right) : \exp -\left(\frac{x}{kT}\right)$$

where

$$x = \frac{e\hbar B}{2mc} = 0.93 \times 10^{-16}\,\text{erg} \quad \text{and} \quad N_{1/2}(N_{-1/2})$$

is the number of electrons with spin parallel (antiparallel) to the field. (Note: The spin and the magnet moment of an electron are opposite in direction since the electron charge is negative.) Therefore the fraction of the Ti irons with their spin parallel to the field is

$$f = \frac{N_{1/2}}{N_{1/2} + N_{-1/2}} = \frac{1}{1 + \exp(-2x/kT)} = \frac{1}{1 + \exp(-1.86/1.38)} = 0.79.$$

(XI-1) According to the uncertainty principle, we know that the angular momentum J of the javelin and its corresponding angular coordinate θ cannot be simultaneously determined to arbitrary degree of precision, i.e.,

$$[I\dot{\theta}_o]\theta_o = \frac{1}{3} m\ell^2 \dot{\theta}_o \theta_o \sim \hbar \quad \text{where } \ell = \text{length of javelin} \tag{1}$$

and θ_o and $\dot{\theta}_o$ are the possible initial angle and angular velocity of the javelin. To estimate the order of the time it takes for the javelin to fall over, we have to estimate the order of magnitude of $\dot{\theta}_o$ and θ_o corresponding to the best balanced position. One simple way is to assume that the javelin is best balanced when the uncertainty in its kinetic energy is of the same order as that in its potential energy. Therefore we obtain

$$\frac{1}{2} \frac{m\ell^2}{3} \dot{\theta}_o^2 \sim \frac{1}{4} mg\theta_o^2 \ell$$

(potential energy taken zero at balanced position) or

$$\sqrt{\frac{\ell}{3}}\,\dot{\theta}_o \sim \sqrt{\frac{g}{2}}\,\theta_o \tag{2}$$

From (1) and (2) we obtain

$$\theta_o^2 \sim \frac{\hbar}{m\ell^2}\sqrt{\frac{6\ell}{g}} \tag{3}$$

and

$$\dot{\theta}_o^2 \sim \frac{3\hbar}{m\ell^2}\sqrt{\frac{3g}{2\ell}} \tag{4}$$

The equation of angular motion is

$$mg \times \frac{\ell \sin\theta}{2} \sim \frac{1}{2} mg\ell\theta \equiv \text{torque} = I\ddot{\theta} = \frac{1}{3} m\ell^2 \ddot{\theta}$$

$$\ddot{\theta} = \frac{3g}{2\ell}\theta .$$

After integration we obtain

$$\theta = \theta_o \exp\left(\sqrt{\frac{3g}{2\ell}}\, t\right) = \sqrt{\frac{\hbar}{m\ell^2}\sqrt{\frac{6\ell}{g}}}\, \exp\left(\sqrt{\frac{3g}{2\ell}}\, t\right) \tag{5}$$

where we have used the initial condition (3). When $\theta \sim 1$, the javelin has practically fallen over. From (5) we obtain

$$t = \sqrt{\frac{2\ell}{3g}}\, \ell n\left(\sqrt{\frac{m\ell^2}{\hbar}\sqrt{\frac{g}{3\ell}}}\right);$$

for $\ell = 1000$ cm and $m = 1$ gm we find $t \sim 30$ sec.

(XI-2)

a. The energy levels of the wave are

$$E = n\hbar\omega = n\hbar Dk^2$$

from which we find the phase velocity of the wave is

$$V = \frac{E}{P}$$

$$= \frac{n\hbar Dk^2}{\hbar k} = nDk = nD\sqrt{\frac{\omega}{D}} = n\sqrt{D\omega}$$

and the group velocity

$$u = \frac{dE}{dP} = \frac{dK}{dP}\frac{dE}{dK} = \frac{1}{\hbar}\frac{dE}{dK} = \frac{1}{\hbar}(2n\hbar Dk)$$

$$= 2nDk = 2n\sqrt{D\omega}$$

b. According to Plank's law, the distribution of the thermal energy is

$$P(E_n) \propto \frac{1}{\exp\frac{E_n}{KT} - 1}$$

from which we obtain the total energy density

$$u = \int_0^\infty E_n P(E_n) \frac{4\pi k^2 dk}{\hbar^3}$$

where $4\pi k^2 dk/\hbar^3$ is the density of the states. Therefore

$$u = n\hbar D \int_0^\infty k^4 \left(\frac{1}{\exp \frac{n\hbar D k^2}{KT} - 1} \right) \frac{4\pi dk}{\hbar^3} = \frac{4\pi n D}{\hbar^2} \int_0^\infty \frac{k^4 dk}{\exp \frac{n\hbar D k^2}{KT} - 1}$$

$$= \frac{4\pi n D}{\hbar^2} \left(\frac{kT}{n\hbar D} \right)^{5/2} f$$

where

$$f = \int_0^\infty \frac{x^4 dx}{\exp (x^2) - 1}$$

is a constant. Therefore $u \propto T^{5/2}$.

(XI-3)

a. The wave function is a function of the space coordinates x, y, z and the time t which describes a system in such a way that

i. it is an amplitude so that it can interfere with itself;

ii. the absolute square of the wave function is proportional to the probability distribution of the system being described;

iii. it describes a single particle (photon, electron, etc.) rather than the statistical distribution of a large number of such quanta.

b.

$$\vec{J} = \vec{J}_1 + \vec{J}_2$$

$$|J| = |J_1| - |J_2|, \; |J_1| - |J_2| + 1, \; \ldots, \; |J_1| + |J_2|$$

where we have assumed $|J_1| > |J_2|$.

c.

$\Delta j = 0, \pm 1$ (but no $j = 0 \rightarrow j = 0$)

$\Delta \ell = \pm 1$

$\Delta m = \pm 1, 0.$

d. The commutator $[G, F] = GF - FG = 0.$

(XI-4) Using the uncertainty principle, we find the momentum of a nucleon confined in a distance Δx is of the order of

$$\Delta P \sim \frac{\hbar}{\Delta x} \sim \frac{10^{-27}}{3 \times 10^{-13}} \sim 3 \times 10^{-15} \text{dyne-sec}$$

where Δx = radius of carbon nucleus $\approx 3 \times 10^{-13}$ cm, from which we find the kinetic energy of the ground state is:

$$3 \times \frac{(\Delta P)^2}{2m} \sim \frac{3\hbar^2}{2(\Delta x)^2 m} \sim 7 \text{ MeV}.$$

The mean energy is expected to be somewhat larger. Alternatively, we can use the expression for the mean nucleon energy according to the Fermi statistics in the case that $kT \ll \epsilon_f$: (See X-4 for derivation of ϵ_f.)

$$<\epsilon> = \frac{3}{5}\epsilon_f = \frac{3}{5}\frac{\hbar^2}{2m}\left(\frac{3\pi^2 n}{V}\right)^{2/3} = \frac{3}{5}\frac{\hbar^2}{2m}\left(\frac{3\pi^2 \times 6}{\frac{4\pi}{3}(3 \times 10^{-13})^3}\right)^{2/3}$$

$$= \frac{3 \times (1.973 \times 10^{-11})^2 \times 12.16}{9380 \times (10^{-13})^2 \times 9} = 17 \text{ MeV}.$$

(XI-5) Number of H_e formed per second equals

$$\frac{4 \times 10^{26}}{26 \times 1.6 \times 10^{-13}} \sim 10^{38} \frac{\text{atoms}}{\text{sec}}.$$

(XI-6) The energy of K X-ray for copper is $13.6 \times (29 - 1)^2$ eV or 10.6 KeV.

(XI-7) Using the 4-momenta equation

$$p_1 + p_2 = p_f,$$

where p_f is the momentum of the final particles, we find the relation

$$2m_e^2c^2 + 2m_eE_e = m_f^2c^2 \quad \text{or} \quad E_e = \frac{m_f^2c^2 - 2m_e^2c^2}{2m_e}$$

The minimum incident energy for the production of an e^+, e^- pair corresponds to the case $m_f = 4m_e$. Therefore $E_e = 7m_ec^2$ or the threshold kinetic energy ≈ 3MeV.

(XI-8) Since the potential energy of the system is minimum when the magnetic dipoles of the unpaired electrons are in the same direction, the spins of the electrons in the $n = 3$ states tend to line up. Therefore $S = (1/2)(6 - 2) = 2$. Furthermore $L = 2$ and $J = 4$ according to Hund's rule. We obtain $^5D_4$, i.e.,

$$(1s^2\ 2s^2\ 2p^6\ 3s^2\ 3p^6\ 4s^2\ 3d^6).$$

(XI-9) The splitting between the two energy levels is

$$\Delta\nu = \frac{1}{h}2\mu_eB = \frac{2e\hbar B}{\hbar 4\pi mc}g_sS = \frac{eB}{2\pi mc} = \frac{e}{mc}\frac{B}{2\pi} = \frac{1.76 \times 10^7 \times 10^4}{6.28} =$$

$$= 3 \times 10^{10}/\text{sec}.$$

(XI-10)

a. A, because $\Delta E_1 = \langle\psi_o, H_1\psi_o\rangle > 0$ for a positive function H_1.

b. A, because $\Delta E_1 = \langle\psi_1, H_1\psi_1\rangle > 0$.

c. D, because $\Delta E_1 = \langle\psi_o, H_1\psi_o\rangle = 0$ for an odd function of H_1, but the second-order perturbation term

$$\Delta E_2 = \frac{|\langle\psi_o, H_1\psi_1\rangle|^2}{E_o - E_1} < 0.$$

(XII-1)

a. Substituting $x = (x_1 - x_2)$, $R = (x_1 + x_2)$, $\psi = \psi_x \psi_R$, and $\mu = M/2$ into the usual Schröedinger equation for two particles of coordinates x_1 and x_2, we obtain the equation corresponding to the relative motion

$$H\psi_x \equiv \left(-\frac{\hbar^2}{2\mu}\frac{d^2}{dx^2} + \frac{1}{2}kx^2\right)\psi_x = E\psi_x$$

where μ is the reduced mass. After differentiation we find

$$\left(-\frac{\mu k}{2\mu}x^2 + \frac{\hbar}{2\mu}\sqrt{\mu k} + \frac{1}{2}kx^2\right)\psi_x = E\psi_x$$

or

$$\frac{\hbar}{2\mu}\sqrt{\mu k}\psi_x = E\psi_x$$

from which we obtain

$$E = \frac{\sqrt{\mu k}}{2\mu}h = \sqrt{\frac{k}{4\mu}}h = \sqrt{\frac{k}{2M}}h$$

b. Let $y^2 = \sqrt{\mu k}\,x^2/\hbar$. Since the wave function of the ground state is

$$\psi_x = \sqrt{\frac{2}{\sqrt{\pi}}}\exp(-y^2/2), \quad \text{we find:}$$

$$\langle P\rangle = |\langle\psi_x, \frac{h}{i}\frac{d}{dx}\psi_x\rangle / \langle\psi_x, \psi_x\rangle|$$

$$= 2\int_0^\infty e^{-y^2} y \frac{(\mu k\hbar^2)^{1/4}}{\sqrt{\pi}} dy = \left(\frac{\sqrt{h\sqrt{\mu k}}}{\sqrt{\pi}}\right)\left(2\int_0^\infty e^{-y^2} y\, dy\right)$$

$$= \frac{\sqrt{h\sqrt{\mu k}}}{\sqrt{\pi}}$$

since

$$\int_0^\infty e^{-y^2}(2y\,dy) = \int_0^\infty e^{-z}dz = 1,$$

and

$$\int_0^\infty e^{-y^2} dy = \frac{\sqrt{\pi}}{2}; \text{ and } \frac{dy}{dx} = (\frac{\mu k}{\hbar^2})^{1/4}.$$

c. Let $\psi(p)$ be the wave function in momentum space. $\psi(p)$ is related to ψ_x by Fourier transformation

$$\psi(p) = \int_0^\infty e^{ipx/\hbar} e^{-\alpha x^2} dx$$

where $\alpha = \sqrt{\mu k}/2\hbar$. We find

$$\psi(p) = \int_0^\infty \exp - (\alpha(x + \frac{ip}{2\hbar\alpha})^2 + \alpha(\frac{ip}{2\hbar\alpha})^2)dx$$

$$\propto \exp (\frac{-p^2}{4\hbar^2 \alpha})$$

$$\propto \exp (- \frac{p^2}{2\hbar\sqrt{\mu k}})$$

The probability of $p < \sqrt{\hbar\sqrt{2Mk}} = p_0$ is

$$\text{prob} = \int_0^{p_0} \psi^2(p)dp / \int_0^\infty \psi^2(p)dp$$

$$= \frac{2}{\sqrt{\pi}} \int_0^{q_o} e^{-q^2} dq \equiv \text{erf}(q_o)$$

where $\text{erf}(q_o)$ is the error function. Using

$$q_o = \frac{p_0}{\sqrt{2\hbar\sqrt{\mu k}}} = \frac{\sqrt{\hbar\sqrt{2Mk}}}{\sqrt{2\hbar\sqrt{\mu k}}} = \frac{\sqrt{\sqrt{2M}}}{\sqrt{2\sqrt{\mu}}} = \frac{\sqrt{\sqrt{4\mu}}}{\sqrt{2\sqrt{\mu}}} = 1.$$

we find prob = erf(1) = 0.84.

(XII-2)

a. Let

$$\psi_{in} = e^{ikz} \text{ and } \psi_{scat} = f(\theta)\frac{e^{ikr}}{r},$$

we find

$$J_{inc} = \langle \psi_{in}, \frac{P_z}{m} \psi_{in} \rangle = \langle \psi_{in}, \frac{\hbar}{im}\frac{d}{dz} \psi_{in} \rangle = \frac{\hbar k}{m}$$

or if we define ψ_{inc} antisymmetric with respect to ψ_{in} and $\bar{\psi}_{in}$, we get

$$J_{inc} = \frac{\hbar}{2im}(\bar{\psi}_{in}(\frac{\partial}{\partial z}\psi_{in}) - (\frac{\partial}{\partial z}\bar{\psi}_{in})\psi_{in})$$

$$= \frac{\hbar}{2im}(e^{-ikz}e^{ikz} \times (ik) - e^{-ikz}e^{ikz} \times (-ik))$$

$$= \frac{\hbar}{2im}(2ik) = \frac{\hbar k}{m},$$

and

$$J_{scat} = \frac{\hbar}{2im}(\bar{\psi}_{out}\frac{\partial}{\partial r}\psi_{out} - (\frac{\partial}{\partial r}\bar{\psi}_{out})\psi_{out})$$

$$= \frac{\hbar}{im}(f^*(\theta)\frac{e^{-ikr}}{r} ikf(\theta)\frac{e^{ikr}}{r})$$

$$= \frac{\hbar k}{m}(\frac{|f(\theta)|}{r})^2.$$

b. From the definition of the differential cross section, it follows that

$$\frac{d\sigma(\theta)}{d\Omega} = \frac{r^2 J_{scat}}{J_{inc}} = |f(\theta)|^2.$$

(XII-3) At 5.5, 6.25, 6.5, 6.75, 7.5; to locate the positions of the peak and the valley of the spectrum as well as where the transition takes place. The asymptotic value, i.e., $E \gg 13$ is also of interest. If the calculation of σ from raw data is simple, we can take three measurements at $E_1 = 5.5$, $E_2 = 6.5$, and $E_3 = 7.5$ at first, and then use the new data to determine the positions of the next two measurements. The idea is to take more data around where the data deviates from a smooth curve. For example, if $\sigma(E_i)$ is considerably different from $(\sigma(E_i + 0.5) + \sigma(E_i - 0.5))/2$, where $i = 1$, or 2, or 3, then we should take data at $E_i \pm 0.25$.

(XII-4) Assuming a Gaussian distribution, we find the probability for detecting n counts is

$$P(n) = \frac{1}{n_{rms}\sqrt{2\pi}} \exp\left(-\frac{(n-N)^2}{2n_{rms}^2}\right)$$

where n_{rms} is the root mean square of the fluctuation in n, and N is the mean counting.

For $N >> 1$, we have $n_{rms} \sim \sqrt{N}$. The probability for $n < 9700$ is

$$P = \frac{1}{n_{rms}\sqrt{2\pi}} \int_0^{9700} \exp\left(-\frac{(n-N)^2}{2n_{rms}^2}\right)dn$$

$$\sim \frac{1}{\sqrt{\pi}} \int_{-\infty}^{x_o} \exp(-x^2)dx \sim 0.0015$$

where

$$x = \frac{n-N}{\sqrt{2}\, n_{rms}} = \frac{n - 10000}{141.4}$$

and

$$x_o = \frac{9700 - N}{\sqrt{2}\, n_{rms}} = \frac{-300}{141.4} = -2.13.$$

(XII-5) The probability of decay at time t is

$$P(t) = \frac{1}{\tau} e^{-t/\tau} = \frac{1}{\tau} \times \text{(the probability for the atom to survive after } t)$$

where τ is given to be ten days. The probability that the atom will decay during the fifth day is

$$\hat{P} = \int_4^5 P(t)dt = \frac{1}{\tau}\int_4^5 e^{-t/\tau}dt$$

$$= -e^{-t/\tau}\Big|_{t=4}^{t=5} = (e^{-0.4} - e^{-0.5}) = (0.670 - 0.606)$$

$$= 0.064.$$

(XII-6) Since the neutron has to be very energetic, its speed must be very close to c. The time t, it takes for the neutron to reach the earth is approximately 10 years or $\pi \times 10^8$ seconds. From the condition that half of them survive at the end of the trip, we find the time passed for the trip measured in a frame moving with the neutron has to equal the half-life time of the neutron, i.e., $t_o = 12$ minutes. Using the relation

$$\gamma \equiv \frac{E_n}{mc^2} = \frac{t}{t_o}$$

where t is the time measured on earth, we find

$$E_n = \gamma(m_n c^2) = \frac{t}{t_o} \times m_n c^2$$

$$= \frac{\pi \times 10^8}{12 \times 60} \times m_n c^2 = (4.4 \times 10^5) m_n c^2 \simeq (4.4 \times 10^5)(940 \text{ MeV})$$

$$\sim 4 \times 10^8 \text{ MeV}.$$

(XII-7) Let $L\hbar$ be the relative angular momentum of the two pions. The parity of the system is $(-1)^L$. Bose statistics requires the two pions to be in an even state, thus L is even for spinless particles like pions. Therefore $S_K = L =$ even.

(XIII-8) On the corners of

(a) an equilateral triangle with each side $= \sqrt{3}$ R

(b) a regular tetrahedron with each side $= \frac{2}{\sqrt{3}}$ R

It is because in these configurations, the charges are separated by the greatest distances, yet are symmetric under the interchange of any two charges.

(XII-9)

$$\frac{Zze^2}{R} = \frac{90 \times 2 \times (4.8 \times 10^{-10})^2}{1.3 \times 10^{-13} \times 6 \times 1.6 \times 10^{-6}} \sim 30 \text{ MeV}$$

where $R \simeq \sqrt[3]{238}\,(1.3 \times 10^{-13}\,\text{cm}) \simeq 6 \times 1.3 \times 10^{-13}\,\text{cm}$.

(XII-10) From the force equation $GMm/r^2 = mv^2/r$, we get

$$GM = v^2 r.$$

Furthermore using the Bohr quantum condition $rmv = n\hbar$, we get

$$v = \frac{n\hbar}{mr}$$

Thus

$$GM = v^2 r = \left[\frac{n\hbar}{mr}\right]^2 r = \frac{n^2\hbar^2}{m^2 r}$$

and

$$r = \frac{n^2\hbar^2}{GMm^2} = \frac{\hbar^2}{GMm^2} = \frac{1}{GMm}\left[\frac{\hbar^2}{m}\right] = \frac{1}{GMm}\left[e^2 a_o\right] = \frac{(e/m)^2}{(GM/m)} a_o \quad \text{if } n = 1.$$

$$a = a_o \frac{e^2/m^2}{GM/m} = a_o \frac{\left(\dfrac{4.8 \times 10^{-10}}{9.1 \times 10^{-28}}\right)^2}{6.7 \times 10^{-8} \times 1840}$$

$$\approx 0.24 \times 10^{40} \times a_o \sim 1.2 \times 10^{31}\,\text{cm}.$$

where $a_o = 5 \times 10^{-9}\,\text{cm}$.

(XII-11) $1/Z$.

(XII-12) $n = 3$, $\ell = 1$, and $n = 2$, $\ell = 1$.

(XII-13)

$$\sum_{\ell=0}^{4} 2(2\ell + 1) = 50.$$

(XII-14) Since the wave function of a fermion-antifermion system must be antisymmetric according to Fermi statistics, we find the following condition

$(-1)^{\ell+s+1} = -1$ or $\ell + s =$ even (See problem VII-3)

must hold. For $\ell = 1$, $s = 1$, i.e., the spins of the electron and the positron are in the same direction. Since the charges of an electron and a positron are of opposite sign, it follows that the magnetic momenta are antiparallel and the resultant is zero.

(XII-15) $\pm 1/2$; $\pm 1/2$; $\pm 1/2$; $3/4$.

(XII-16) $\pm \frac{e\hbar}{2mc} B$.

(XII-17) $p = mc/2$.

(XII-18) $h\nu >> \alpha^2 m_e c^2$ where the binding energy is $1/2\, \alpha^2 m_e c^2 = 13.6$ eV.

(XII-19) $\ell = 1$ in quantum mechanics, which is the least symmetric state; $\ell = 0$ in semiclassical mechanics (e.g., in Sommerfeld's model), which describes a circular orbit. In classical mechanics we know the angular momentum $\vec{\ell} = \vec{r} \times \vec{p}$ is maximum when $\vec{p} \perp \vec{r}$, i.e., for a circular motion.

(XII-20) 4×13.6 eV.

(XII-21) Compton scattering and photoelectric effect.

APPENDIX

Table I Some Useful Physical Constants

N_o	$= 6.02217 \times 10^{23}$ mole^{-1}
e	$= 4.8033 \times 10^{-10}$ esu $= 1.6022 \times 10^{-19}$ coulomb
1 MeV	$= 1.6022 \times 10^{-6}$ erg
$\hbar$	$= 6.5822 \times 10^{-22}$ MeV-sec
	$= 1.05459 \times 10^{-27}$ erg-sec
$\hbar c$	$= 1.973 \times 10^{-11}$ MeV-cm = 1973 ev Å
α	$= e^2/\hbar c = 1/137.036$
$k_{Boltzmann}$	$= 1.3806 \times 10^{-16}$ erg-K^{-1}
	$= 8.617 \times 10^{-11}$ MeV-K^{-1}
m_e	$= 0.511$ MeV
m_p	$= 938.26$ MeV $= 1836.1\ m_e$
μ_{Bohr}	$= e\hbar/2m_e c = 0.57884 \times 10^{-14}$ MeV-gauss^{-1}
$\mu_{nucleon}$	$= e\hbar/2m_p c = 3.1525 \times 10^{-18}$ MeV-gauss^{-1}
Acceleration by gravity	$= 980.62$ cm-sec^{-2} at sea level, 45°
Gravitational constant	$= 6.6732 \times 10^{-8}$ cm^3-g^{-1}-sec^{-2}
1 calorie	= 4.184 joules
1 atmosphere	$= 1033.2275$ gm-cm^{-2}

Table II <u>Conversion Table between mks and Gaussian units</u>

Physical Quantity	Symbol	Rationalized mks	Gaussian
Length	ℓ, d, r	1 meter (m)	10^2 centimeters (cm)
Mass	m	1 kilogram (kg)	10^3 grams (gm)
Time	t	1 second (sec)	1 second (sec)
Force	F	1 newton	10^5 dynes
Work Energy	W E, U	1 joule	10^7 ergs
Power	P	1 watt	10^7 ergs-sec^{-1}
Charge	q	1 coulomb (coul)	3×10^9 statcoulombs
Charge density	ρ	1 coul m^{-3}	3×10^3 statcoul-cm^{-3}
Current	I	1 ampere (coul sec^{-1})	3×10^9 statamperes
Current density	J	1 amp m^{-2}	3×10^5 statamp-cm^{-2}
Electric field	E	1 volt m^{-1}	$\frac{1}{3} \times 10^{-4}$ statvolt-cm^{-1}
Potential	V	1 volt	$\frac{1}{300}$ statvolt
Polarization	P	1 coul m^{-2}	3×10^5 statcoul-cm^{-2} (statvolt cm^{-1})
Displacement	D	1 coul m^{-2}	$12\pi \times 10^5$ statvolt-cm^{-1} (statcoul-cm^{-2})
Conductivity	σ	1 mho m^{-1}	9×10^9 sec^{-1}
Resistance	R	1 ohm	$\frac{1}{9} \times 10^{-11}$ sec-cm^{-1}
Capacitance	C	1 farad	9×10^{11} cm
Magnetic flux	φ	1 weber	10^8 gauss-cm^2 or maxwells

Physical Quantity	Symbol	Rationalized mks	Gaussian
Magnetic induction	B	1 weber m^{-2}	10^4 gauss
Magnetic field	H	1 ampere-turn m^{-1}	$4\pi \times 10^{-3}$ oersted
Magnetization	M	1 weber m^{-2}	$\frac{1}{4\pi} \times 10^4$ gauss
Inductance	L	1 henry	$\frac{1}{9} \times 10^{-11}$

REFERENCES

Books are listed in the order of increasing sophistication under each catagory.

I. Mechanics:

French, A. P., Newtonian Mechanics, New York: W. W. Norton, 1971.

Ford, K. W., Classical and Modern Physics, Vol. I, Stamford, Connecticutt: Xerox, 1972

Sears, F. W., Mechanics, Heat, and Sound, Reading, Massachusetts: Addison-Wesley, 1950.

Berkely Physics Course, Vol. I, Mechanics, New York: McGraw-Hill, 1965.

Symon, K. R., Mechanics, Reading, Massachusetts: Addison-Wesley, 1971.

Goldstein, H., Classical Mechanics, Reading, Massachusetts: Addison-Wesley, 1950.

II. Relativity:

Taylor, E. F. and J. A. Wheeler, Spacetime Physics, San Francisco: W. H. Freeman, 1966.

III. Electricity and Magnetism:

Ford, K. W., Classical and Modern Physics, Stamford, Connecticutt: Xerox, 1972.

Halliday, D. and R. Resnick, Physics, Part 2, New York: Wiley, 1966.

Berkeley Physics Course, Vol. II, Electricity and Magnetism, New York: McGraw-Hill, 1965.

Hauser, W., Introduction to Principles of Electricity, Reading, Massachusetts: Addison-Wesley, 1971.

Panofsky, W. K. and Phillips, M., Classical Electricity and Magnetism, Reading, Massachusetts: Addison-Wesley, 1962.

Jackson, J. D., Classical Electrodynamics, New York: Wiley, 1962.

IV. Heat:

Zemansky, M. W., Heat and Thermodynamics: An Intermediate Textbook, New York: McGraw-Hill, 1968.

Morse, P. M., Thermal Physics, Menlo Park, California: W. A. Benjamin, 1969.

V. Optics:

Jenkins, F. A. and H. E. White, Fundamentals of Optics, New York: McGraw-Hill, 1957.

Rossi, B., Optics, Reading, Massachusetts: Addison-Wesley, 1957.

VI. Statistical Mechanics:

Berkeley Physics Course, Vol. V, Statistical Mechanics, New York: McGraw-Hill, 1967.

Tolman, R. C., Principles of Statistical Mechanics, New York: Oxford University Press, 1938.

Huang, K., Statistical Mechanics, New York, Wiley, 1963.

VII. Solid State Physics:

Kittel, C., Elementary Statistical Physics, New York: Wiley, 1958.

Kittel, C., Introduction to Solid State Physics, New York: Wiley, 1971.

VIII. Quantum Mechanics:

Saxon, D. S., Elementary Quantum Mechanics, San Francisco: Holden-Day, 1968.

Merzbacher, E., Quantum Mechanics, New York: Wiley, 1970.

Bohm, D., Quantum Theory, Englewood Cliffs, New Jersey: Prentice-Hall, 1951.

Schiff, L. I., Quantum Mechanics, New York: McGraw-Hill, 1968.

Messiah, A., Quantum Mechanics, New York: Wiley, 1961-1962.

Bethe, H. A. and R. W. Jackiw, Intermediate Quantum Mechanics, Menlo Park, California: W. A. Benjamin, 1968.

IX. Modern Physics:

Acosta, V. et al, Essentials of Modern Physics, New York: Harper & Row, 1973.

Eisberg, R. M., Fundamentals of Modern Physics, New York: Wiley, 1961.

Harnwell, G. P. and W. E. Stephens, Atomic Physics: An Atomic Description of Physical Phenomena, New York: Dover, 1966.

Reimann, A. L., Physics, New York: Harper & Row, 1971.

Born, M., Atomic Physics, Hafner & Blackie, 1969.

Fermi, E., Nuclear Physics, Chicago: University of Chicago Press.